Advance Praise for *Sunshine State* by Sarah Gerard

"Gerard is a virtuoso of language, which in her hands is precise, unlabored, and quietly wrought with emotion. . . . She is also a very diligent journalist. . . . Brave, keenly observational, and humanitarian. . . . Gerard's collection leaves an indelible impression."

—*Publishers Weekly*, starred review

"These large-hearted, meticulous essays offer an uncanny X-ray of our national psyche, examining that American mess of saints and con men, the peculiar, culpable innocence that confuses money and moral worth, charity and personal aggrandizement. Gerard's prose is lacerating and compassionate at once, showing us both the grand beauty of our American dreams and the heartbreaking devastation they wreak."

—Garth Greenwell, author of *What Belongs to You*

"Sarah Gerard's *Sunshine State* gloriously gutted me—and by that I mean changed me forever as a reader. Using Florida as a lens and the body as a ticket to travel, Gerard weaves her astonishing prose through land and corporeal truth, like the inside out of love and loss and violence and beauty. What if our obsessions and addictions and longings were actually knotted up in our guts with environmental destruction, religion, and some insane storyline called love? What if our bodies carry the trace evidence of all the places we've inhabited? What if our relationships were laid bare like a beach? *Sunshine State* reminds us of who we really are underneath the skin we live in and the ground we stand on—and mercifully, there is still beauty, in spite of everything."

—Lidia Yuknavitch, author of *The Book of Joan*
and *The Chronology of Water*

"Good writers learn early in their careers that what people are most interested in is other people. In this regard, Sarah Gerard is indeed a very good writer. Her prose sparkles in this series of essays, but it is the people in *Sunshine State* who capture and concern us. Vivid, sometimes disturbing, but always [illegible] r of our southernmost state where a [illegible] rug-gle, and die beneath tropical sk [illegible]

—Homer Hickam, author *Car* [illegible] *oys*

"With visceral wit and a literary toolkit full to the brim with new forms, Sarah Gerard's first collection of essays makes the wild and untamed inner life of Florida bloom vividly within the reader's mind. *Sunshine State* is a strange, thoughtful, and deeply felt journey through a state whose beauty and peril speak to the contradictions of an entire nation."

—Alexandra Kleeman, author of *Intimations and You Too Can Have a Body Like Mine*

"I've never read anything like Sarah Gerard's *Sunshine State*, and I'm worse off as a writer for it. Gerard manages to personalize the political and politicize the personal in ways that feel at once effortless and insanely ambitious. These essays remind me that at the bottom and top of structural inequities are people with emotions, friends, failures, and memories of love and loss. *Sunshine State* is a book of essays, but really it defies and embraces form. That formal defiance leads to some of the best essays I've read in the twenty-first century. *Sunshine State* should be mandatory reading for everyone living in Florida, the United States, and the world. It's an amazing creation."

—Kiese Laymon, author of *Long Division* and *How to Slowly Kill Yourself and Others in America*

"Sarah Gerard writes with soulful clarity and keen intelligence about the cultish relationships and aspirational thinking that course through American life. This is a collection packed with bittersweet longing—for a life that's fuller or wilder or wealthier, for a larger self that's always out of reach."

—Alex Mar, author of *Witches of America*

"Sarah Gerard's sparkling essays-as-memoir is as multifaceted as Florida itself. Navigating intense friendships; her family's unconventional faith; a flirtation with Amway, tattoos, drugs, boyfriends, and a husband; a homeless shelter and a bird sanctuary run by a corrupt madman, Gerhard is wide-eyed yet fully present, blunt yet empathetic to not only the crazy swirl of characters that surround her, but to herself in formation. A tough, honest, beautiful work by one of our brightest and most unflinching young writers."

—Rob Spillman, editor of *Tin House* and author of *All Tomorrow's Parties*

"As a fellow Floridian, it was a great gift to discover such a viscerally refreshing vision of home in Sarah Gerard's *Sunshine State*. Gerard masterfully explores the environmental, economic, and regional complexities of Florida alongside the eternal mysteries of identity, home, family, trauma, and desire. A stellar essay collection by a writer in possession of a talent as singular and furious as Florida itself."

—Laura van den Berg, author of *Find Me*

"For those who fear Florida is comprised primarily of gators and the insane, this book may seem like it was written for you. In many ways, it surely was, giving life and voice to a world which has previously not held much acreage in your mind. But at its core, *Sunshine State* is a love letter to the wild and fascinating land itself, and the cast of characters who call it home."

—Amelia Gray, author of *Isadora* and *Gutshot*

"In *Sunshine State*, Gerard goes deep into the paradoxes of her birth state. I found these essays to be smart, kind, and illuminating. This book left me improved spiritually."

—Darcey Steinke, author of
Sister Golden Hair and *Suicide Blonde*

"Intensely personal and intricately researched, Sarah Gerard's essays break ground with the work of Eula Biss, Maggie Nelson, Joan Didion. Gerard is provocative and an excellent sleuth. She digs for the secret, unshakeable truths we are busy turning away from—yet she is never sensational, never sentimental. Her mind is tough but she reaches with love. She asks that we reach with her—with her resilience, her prodigious strength. This book is a gift to all of us."

—Noy Holland, author of
I Was Trying to Describe What it Feels Like and *Bird*

"Sarah Gerard has that lingering gaze shared by investigative journalists and lovers. With equal measures of scrutiny and tenderness, she examines her feverish homeland and its denizens, herself and those she loves. No idealization can withstand this kind of scrutiny, and thank God, because I would not trade the hours I've spent in Gerard's world for any more perfect version. One can't hope for a more sharp-eyed, tender-hearted chronicler of herself and our busted world."

—Melissa Febos, author of *Whip Smart* and
Abandon Me

"Sarah Gerard's *Sunshine State* weaves narrative nonfiction and personal essay in a way that creates its own unique tapestry. This essay collection is unlike any other I've encountered—stylistically dazzling without sacrificing a reporter's precision, relentlessly moving without doling a sentimentalist's artificial sugar—this is a collection of so many Floridas only a native could know. At a time when America feels so broken, Gerard allowed me to love it again somehow."

—Porochista Khakpour, author of
Sick and *The Last Illusion*

"Brilliant, empathetic, fearless, and humane—in her search to better understand herself, her family, and the state that helped shape her, Sarah's insight, heart, and diligence are boundless. The best book of essays I've read in years—a brilliant collection from a writer of incredible versatility and talent."

—J. Ryan Stradal, author of
Kitchens of the Great Midwest

"Sarah Gerard's *Sunshine State* is a deeply intelligent, personal, and political collection of rich essays, with a clarity sharp as an icicle and "place" as the connective tissue. The themes of class, identity politics, and loneliness emerge in ways that are simultaneously disturbing and comforting. The perfect book for the complex and heady humans in your life—aka, for everyone."

—Chloë Caldwell, author of
Women and *I'll Tell You in Person*

"Armed with a mesmerizing breadth of empathy and a rare, hi-res emotional intuition, Sarah Gerard's essays lead us forward through decades of her observation of our world, along the way unpacking everything from religion to economics, desire to aspiration, grief to the very grit of what seems to make a person tick. It's rare to find a voice you can come to believe in so quickly and completely, like an old friend, and one whose very spirit makes the world seem that much more bearable, more true. Here is something to believe in."

—Blake Butler, author of *300,000,000*

"Sarah Gerard's *Sunshine State* probes at the fringes of society, the intersection of right and wrong, the private core of our fundamental self-definitions. Sarah's compassionate and boundlessly curious essay collection drives always toward truth, even when that truth is hard to bear. An unforgettable book, by a writer with a powerful, essential American voice."

—Julie Buntin, author of *Marlena*

"Compelling, intriguing, and brimming with voice, *Sunshine State* showcases Sarah Gerard's huge heart and huge brain. Each of her essays is a relentless interrogation—of herself and of the world around her. She is a seeker and a seer, a critic and an empath, an intellectual and a poet. This book isn't just about Florida; it's about America. It's about humanity. I could read Sarah Gerard all day."

—Diana Spechler, author of
Who by Fire and *Skinny*

"Sarah Gerard brings an immersion journalist's acuity and shrewdness to essays made urgent by a native daughter's alloy of sympathy and rage. Capacious and captivating, *Sunshine State* gets Florida right—and dead to rights—while breathing fresh life into the shoe-leather memoir."

—Justin Taylor, author of *Flings*

"Sarah Gerard's writing is so precise, so deft, so marvelously human, so deeply connected to the people around her, that if I were to have my choice of executioners, I'd call on her."

—John Reed, author of *Snowball's Chance*

"An insightful look into the darker corners of our sunny paradise from a skilled writer and sharp-eyed observer, *Sunshine State* is a collection of essays that weaves a portrait of Florida as a place that's closer to the police blotter reality than the puffery of the tourism promoters."

—Craig Pittman, author of *Oh, Florida!*

SUNSHINE STATE

ALSO BY SARAH GERARD

Binary Star

SUNSHINE STATE

ESSAYS

SARAH GERARD

HARPER PERENNIAL

NEW YORK • LONDON • TORONTO • SYDNEY • NEW DELHI • AUCKLAND

The following essays previously appeared elsewhere in slightly altered forms: "BFF" was originally published as a chapbook by *Guillotine*; "Going Diamond" was originally published in *Granta*.

This is a work of nonfiction. Except where indicated otherwise, the events and experiences detailed herein are all true and have been faithfully rendered as remembered by the author, to the best of her ability, or as told to the author by those who were present. Some names, physical descriptions, and other identifying characteristics have been changed to protect the privacy and anonymity of the individuals involved.

HarperCollins books may be purchased for educational, business, or sales promotional use. For information, please email the Special Markets Department at SPsales@harpercollins.com.

FIRST EDITION

Designed by Leydiana Rodriguez

Library of Congress Cataloging-in-Publication Data has been applied for.

ISBN 978-0-06-243487-6 (pbk.)

17 18 19 20 21 LSC 10 9 8 7 6 5 4 3 2

To Mother-Father

The Florida sun seems not much a single ball overhead but a set of klieg lights that pursue you everywhere with an even white illumination.

—John Updike, *Rabbit at Rest*

Life feeds on life, it has no choice. Life stokes its furnace with life, but not with life in general; it burns the unique and particular life of the individual, and when there's nothing left to feed to the flames, the fire goes out.

—César Aira, *Shantytown*

CONTENTS

SUNSHINE STATE

BFF

I'll begin our story with that afternoon, we hadn't spoken for a year—like so many years when we didn't speak—when you pulled up next to me on my walk to work and offered me a ride. I climbed into the passenger seat of your Dodge Omni, knowing I'd begun another cycle, and still another. Feeling that I'd been tricked again, that I should have refused. You caught me in a moment of not knowing.

If I start there, I should say that by this time we'd already gotten the tattoos that linked our right and left hips together into a single message: "Forever / & ever." And I should say that, at a glance, my text appeared to spell "Beaver"—too perfect that yours bore the autonomous word while mine was dependent.

Do you know mine is covered up? I hope yours remains. I hope you still see it in the mirror and it surprises you some mornings in your half sleep. You're the only woman I've loved this way: enough to want to hurt you.

I'm reminded of your first tattoo, on your back (tramp stamp, so Florida). You told me you orgasmed while getting it, your nerve endings so close to the surface of your skin. Like many things you told me, I don't know if it's true. Even twenty years after we met, I can't tell when you're lying, but have learned to assume you often are. I've learned to hope you are sometimes.

Our friendship was a sticky web. Our friendship was a black box. Our friendship was a swamp full of cottonmouths.

Another of your tattoos: an alligator eating up your left arm. (Is it left, like mine, or right? Are we mirrors even now?) The tattoo covers another, some text you got when I got my text. Mine: "Resilient in the clutch of darkness." Yours: "Your blues ain't like my blues." Whose pain is deeper? Who wins?

I have to start over, like we started over so many times, with the call you made in our nineteenth summer. We hadn't spoken for a year. I went away to college in New York; you stayed in Florida. You didn't go to school. Nobody in your family went to school.

If I could quantify what came between us: the cost of tuition.

If I could quantify what came between us: the cost of your mother's one-bedroom rent subtracted from my parents' four-bedroom house in a gated neighborhood.

If I could quantify what came between us: the cost of diapers, formula, hospital bills.

The last time we spoke before this, we planned to get coffee and you never called me back. You never showed; you disappeared. That day, I wore a pink slip I'd turned into a dress, copycatting your style, which has always been effortless.

You shinier. You prettier. You taller. You thinner, more popular.

In middle school, you had friends and I had you. You made it so you were my only friend. Don't think I didn't know.

Was it really so important, so often, to know you were first? You were first.

It's stupid I'm still mad at you for this. It's stupid I'm even still mad at you for things you did when we were ten. When you told me you had gone on a date with my boyfriend, he was hardly my boyfriend.

And what was I wearing on the day you finally called, a full year later, linking summer to summer? I don't know now. But I was driving that same car some thousand miles away from you, and you asked me to spell your daughter's name. I got it wrong and you laughed.

Wasn't I supposed to be an English major? Yes, but I'd never heard that name. It's unusual, like yours. It means "pure" in the language of your ancestors. In another it means "sea," like the gulf we grew up swimming in together. The many, many ways we're the same: blue eyes, brown hair, small breasts, freckles, and the moles on our bellies, which we called the moon and stars. Your daughter has your eyes.

Your body: You were always biting your fingernails. Your fingers are long and thin, like your arms, like your legs, like your nose. Do you remember saying my nose was a ski slope? You traced your finger down and it leapt into the air, and you laughed your loud, bigmouthed laugh.

Your body: Years later, you asked me why I didn't tell you that you were too good for stripping. I didn't know you wanted me to tell you.

My body: Do you remember saying my shoulders were broad, that I couldn't wear the tank top we'd just bought together, two of the same, and how I didn't wear it? But you did.

My body: Or the afternoon when you got on my bus without permission and we walked home through a tornado? I wore a black flowered skort with a matching tank top and leather platforms. We took shelter in the covered walkway of a Baptist church and shivered and laughed as the rain painted everything white.

Your body: Or the funny face you made by pinching your nose closed and inhaling hard, crossing your eyes. Or the face you made when I borrowed your pants and laughed so hard I wet them in the back of our friend's dad's car coming back from an eighth-grade party.

Yours: You were nearly paralyzed giving birth the first time. You didn't want an epidural, but the nurse convinced you and then botched the administration. You told her something was wrong, but she didn't believe you. You told me this many years later.

Yours: You dreamt of being a model.

Mine: I split my face open jumping from a train I hopped with the crew change you gave me. I'd never model. Later, I embellished my tattooed text with a half-sleeve train emerging from a tunnel.

Mine: You said my ass was big, and I still believe you.

Ours: We slept with our hips together, holding hands.

The year we met, my parents moved my family from the house where you first knew me, where my father answered the door our first time together and saw your mother standing on the other side of the screen and called her by name. We joked that they had been lovers. We joked or we wished.

You were the closest thing I had to a sister. I knew your body like it was my body; I knew it as it was changing. I knew when you got your period before me because I could smell it. And because you told me about it, drew me pictures of it, told me everything you could about how the blood clotted, how it felt, how it tasted. I knew your breasts as they were changing into shapes I could hold in my hands. I saw your naked body so close to mine so many times in the daylight, and the darkness, and the water, and the moon. I saw you pee. I saw your pubic hair changing color. I saw your face getting older.

I loved you when you grabbed me by the sides of my head and smelled my hair. I do that to my husband now. You do that to your husband now, and to your children. You loved me like a child and a sister and a mother. You were my twin.

I loved you when you held my hand expecting that I would hold yours back, and I did.

We ran down the streets of my neighborhood, prepubescent, wearing nothing but skin, hiding behind trash cans when trucks turned corners and caught us in spotlights. We climbed into other people's boats suspended from docks over dark water, hearing fish below hurl themselves into air, smacking down on the surface.

I lost my virginity so soon after you because of you. You were the first one I told. I wrote you a letter the day it happened and drew you pictures, wanting you to see. My mother found it and put me on the pill. I lied and said that I loved him. He was just some boy.

When your mother finally earned enough to move you into the condo, we held on to the sides of your hot tub and fucked the jets with our knees above the water. It was there that you met your daughter's father, lying by the side of the pool in the sticky sun, turning brown the way I never could. You wanted to catch his eye and it worked.

You knew you were pregnant as soon as you stood up from the grass where you lay with your daughter's father. I see you walking your bicycles back to the duplex you shared with him then, with its haunted Murphy bed and its daisy-colored sunroom—the fear in your face and the hexagonal sidewalk sections beneath you, and this is something you've said to me that I believe.

YOUR MOTHER SLID DOWN THE LAUNDRY CHUTE OF A FOSTER HOME ON SHEETS OF WAXED paper. She described the sound of climbing back up: *boom—boom.*

We were teenagers when she gave you to the state because you couldn't be trusted, with her or with yourself.

There were times I knew you hated her. You'd come home to her on the recliner, corpse-like: TV on the floor, drink in hand, Misty cigarettes on the particleboard table, overweight, hair dyed red, made-up, drunk. She called me to complain about you. Did you know that? She took your daughter when you were coked up and homeless and could no longer protect her. You couldn't even protect yourself. I think of the midnight you said you were going to break into your abandoned childhood home. It was somewhere to sleep.

When you lived in that house your bedroom was a converted living room. A door led from your room to the front of the house; you both entered from the back. A boy lived in your neighborhood: your first love, the first male since your father to touch your bare skin. You were seven in the attic of another boy's house. There were two others there. You told me this story.

And nearby, the day we went exploring in the ruins of an abandoned house, I leapt from one wooden plank to another and landed on a rusty nail and cried. We were ten. I was wearing white Keds and bobby socks. By the time we got back to your house, the heel in my shoe was red and I was leaning on you, heavy. Your mother shook her head as she called my mother.

I remember your house was blue.

You moved out of it the year after I met you, into the low-rent apartment across from the Bosnian children you found annoying even though they were refugees from genocide. You came home alone after school and ate cheese quesadillas from the microwave, a survival technique I found strange until later, when I first heard you say "poor."

You turned twelve. At your party a man passed us and returned to where we chased one another on the grass. We hid from him behind a vined white lattice, giggling at the fact of his maleness. You broke loose from us, screaming, waving something above your head—a sunhat? You knew to be afraid.

I walked as far down the driveway as my parents' cordless phone allowed me that night when you said you were going away. All the houses slept with their eyes open, and the bush by our garage smelled of jasmine, and I was trying to cry, but I couldn't. So I had to pretend.

I pretended other times with you. I pretended we were still best friends when we got our tattoos. I pretended to take your side when my father fired you after giving you a job when you were desperate. I pretended sympathy when you told me you wanted to kill yourself. Your daughter was three; how could you? I pretended to trust you many times when I didn't. You shouldn't have trusted me, either.

What you were wearing the day we came to pick you up at the girls' home, my parents and I, and nobody asked us our names, or made us sign things, or even saw you leaving: A coral-colored vintage dress cinched at the waist. A necklace that hung between your breasts. You had gained weight from the Paxil and you called the dress your Marilyn. You wore, maybe, plastic flip-flops and anklets. Short hair. And I wondered if you wished they cared more, the people at the home. I wondered who had cared about you other than me. You can love someone without trusting her—you know this better than I do. I loved you completely. I love how you think water can feel things. I love that you always smell of sage. I love that you découpaged your coffee table. I love how you think you're punk rock, even now.

You were diagnosed, but you were never crazy. You were caught between the fear of leaving your mother and the fear of staying. Caught in the horror of being complicit in your endangerment every time she told you to fix her a drink. Caught in the need to protect her, though she didn't protect you; caught between the habit of being a child, obedient, and the need to rebel, as a woman.

It's not your fault you couldn't escape.

There was so much I didn't know about you, and I'm angry with you for thinking that I did. I'm angry with myself for failing to see it. For being naïve for so long.

I hurt you.

Twice you ran to Miami: First, when you fled that home with another girl, who you told me looked like a pixie. I wonder if that was true. Second, when you awoke in a hotel shower, covered in blood, with your daughter half a state away and no way to escape the room.

I hope that was a lie.

Lies I know you told me:

You were abducted by aliens in the middle of the night when you were five. They took you from the kitchen in a bright light that came down from the ceiling. Your mother found you sleeping on the tiles in the morning. You never told her about the abduction.

You modeled for a JNCO ad when we were thirteen. In the ad, you were standing on a glass floor looking down into the camera. The cuff of the jeans was twenty-one inches in diameter, the biggest JNCO style to date.

You are distantly related to Chris O'Donnell, whose last name is a derivative of your family name. Or, more likely, your last name is a derivative of his family name. You met him once.

You used JTT's bathroom.

You were writing the dialogue for a pilot cartoon with a famous producer who specifically asked you. This person was a good friend of yours who could help me with an after-school arts program I was trying to start at the time. You gave me his email address. The email bounced back.

You were dating a member of Bob Marley's family. You left your daughter with your mother to go to Miami to meet him at a music festival. It is because of him that you awoke in the hotel shower, covered in blood. He abandoned you there.

You weren't sleeping with the man who worked for my father at the same time you did and who borrowed your car to visit his mother in Tampa. He got high on heroin on the way and totaled your car. You couldn't get to work.

You know things about the "sacred art" of tai chi.

In New York, you accidentally dialed your boyfriend, now the father of your second child, and let him listen to a conversation we were having about a man you wanted to fuck. When he called you back screaming, you were surprised.

You didn't know, when you convinced your stepfather to buy you and your daughter tickets to California, that you were going to find your way from San Diego to LA to be at my courthouse wedding.

You weren't a little proud when our friend picked you to give her solace after I cheated on her friend with the man I later married. You invited her over and listened while she told you all my secrets. You weren't a little excited when you sat down to write me the email that ended our friendship. It didn't thrill you to feel superior in that moment, knowing that I could also lie.

You wilder. You freer. You louder. You bolder. You bigger. More jaded.

To try to set limits was a betrayal. There was always something you were asking for that I wasn't prepared to give. Your stepfather wasn't even your stepfather anymore when you arrived at his San

Diego mansion with your daughter, already pregnant with your second. You told me you found him full of pills and booze, that you couldn't stay there. You called me in LA and I told you to come. What choice did I have?

In the hours before my wedding, we hiked to the top of Barnsdall Art Park, overlooking the city. We ate figs on the grass with my almost husband, and your daughter disappeared into the trees to peer through the windows of the quieted art museum. At the wedding, she posed between us for a family photo: a child to three, all of us smiling.

I made the mistake of needing you once, too. Before you visited me in New York, I told you I had a choice to make and asked you to help me make it. I'd fallen in love with a man who wasn't my boyfriend. You wanted this to be our time and instead I made it mine.

Were you mad at me for betraying a man who hadn't abused me?

Or were you mad at your father, who choked your mother while you watched when you were three? At the man who stalked you to work at the strip club and threw plates at you while your daughter slept in the next room, whose house you fled for a battered women's shelter. The man who overtook you at a party, whom all of your friends still talked to. Your daughter's father, whose name you wouldn't allow to appear on her birth certificate, who later went to prison.

Or your then-boyfriend, who once swung the broad side of his shovel into your pelvis; who came home drunk one night and peed on you while you slept; who dragged you across your apartment by your hair; who, you once explained, you find sexy because he's primal.

After the wedding, you wanted to ride with us to Malibu to throw our vows into the water. I said no. You weren't invited.

When you found out you were pregnant a second time you

asked me if you should keep the baby. I sat on the edge of my ex-boyfriend's bed in upper Manhattan, watching him smoke out the window with his ear cocked, and I told you to do what you felt was best.

I should have known you were asking not only about the baby but also about her father. It felt like a test, like the night you brought me to Mermaids to watch you do tricks on the pole, and I was supposed to say it was beneath you. I should have known it was just a performance. Instead, I was sitting in the audience of another school play, hearing you forget the lines. But you told me once that nothing is worse than being told you fucked up when you already know.

Later, you asked me to spell her name, too: Should there be an *H* or not? I didn't answer that question. I knew you were asking if I'd be there for her.

The first of the last tests.

I'm sorry.

Lies I heard you tell yourself:

Vitamins from fruits and vegetables are concentrated most densely in the stems. Eating the raw stem of a zucchini is more beneficial to your health than eating all the other parts, cooked or uncooked.

The earth is hollow. There is a hole in Antarctica, or the North Pole, depending on the story. It is visible from space, as proven by photos you've found online. A navy admiral named Richard E. Byrd once landed his plane near the hole and encountered a race of humanoids living inside the planet in a land called Agartha. This race has witnessed all the events of human history. They brought Byrd inside the planet and sent him back to civilization with the message that we're headed for extinction.

Present-day humans are the descendants of a technologically advanced extraterrestrial race called the Anunnaki who came to Earth from the planet Nibiru and bred with *Homo erectus*. We know this because a man named Zecharia Sitchin has translated ancient Babylonian tablets and uncovered details of Sumerian myth, which also prove that Ur was destroyed by nuclear fallout from the war between extraterrestrial factions.

The government controls our minds with chemical-filled contrails dragged across the sky behind airplanes. As a society, we're growing progressively stupider and more complacent from breathing in these chemtrails, which can also control the weather.

Everything is controlled by the Illuminati. If you look closely enough, you can find evidence of it anywhere: in the Gmail symbol, in a skirt Beyoncé wore to an awards show, in the dollar bill, and even in the Florida state flag. They are responsible for the rises to power of several important people as well as for countless suspicious deaths. We are all their slaves.

Some people are psychic, and others have special insight into the stars. When you were a child, your mother took you to visit her psychic, who became a close family friend. I went with you once, but she didn't tell my fortune.

Tarot cards can tell you things about the future.

You and I are cosmic twins. Each night, while we're sleeping, we find each other, no matter where we are. No matter if I'm in New York. No matter if you're in Florida. We have been together since the dawn of the universe in infinite forms. We were once an Indian raja and his daughter.

You can force people to love you. If you doubt their commitment, you can force them to prove themselves. You can do this to them over and over, and you will be justified, even when they say they're tired.

The love of a friendship should be limitless. You should be prepared to do anything for each other.

Lies I told you:

A ghost grabbed my toe in the dark one of many nights you slept at my house. We lay awake for hours awaiting another toe grab that never came.

I used JTT's bathroom.

I was sleeping the night before I moved to New York with my then-boyfriend, while you stood outside his apartment, calling and calling, and I didn't pick up. You and I had plans to sing karaoke. He and I were actually fucking.

My mother liked you.

I was a fan of Johnny Cash when I slipped his *Greatest Hits* into the CD player as we crossed the bridge to the beach one night when I was home from college.

I would read *Women Who Run with the Wolves*. I never intended to read it. At first, I didn't trust your recommendation. Now it reminds me too much of you.

I have never looked down on you.

I have helped you at all times without judgment.

When I said I'd be your daughter's godmother, I was ready.

When you called the bookstore where I worked on your birthday, a month after the last time we stopped speaking, you were told I wasn't there. You called me by my nickname and told the bookseller who answered the phone which section I managed. You called yourself a name other than your own, to try to trick them. I'd told them to lie.

Your daughter emailed me from your account a year after we stopped speaking. I responded with pictures and a list of accomplishments, knowing you'd be jealous.

Driving through downtown one night past the Detroit Hotel, you told me you'd been reading Schopenhauer's *Essays and Aphorisms*. I insisted you had misunderstood the title, but I was just an idiot.

I don't wonder what I'd say if I saw you again in passing. When I visit my parents in Florida, I don't make a point to go downtown in the hope that you'll be there. You'd be eating pizza on the sidewalk outside Fortunato's, surrounded by friends. You'd be smiling and confident, the loudest and most interesting of the group. You would talk with your mouth full, like you always have, and would clear the food from your gums with a nail-bitten finger. I would stand across the street, watching you silently in the dark. Your daughter would see me first. I would smile and wave to her.

The detritus of a happy friendship: the tin Beatles poster; your chest of special things, including your copy of a newspaper article about Robert E. Lee, your great-great-great-uncle; the notes we tossed back and forth in Spanish class (Señora Santiago hated us); your journals, which still sit atop my refrigerator; the copy of Khalil Gibran's *The Prophet* that you gave me in New York, an inscription inside from one early-century friend to another; the Bhagavad Gita; the glass pelican sun ornament your daughter gave me, now broken; matching red parachute jackets; matching sneakers and haircuts; No Doubt's *Tragic Kingdom*; the photograph I keep turned backward in a frame engraved with "Best Friends": we're posing for each other.

I know you live on a peach orchard now. Last year, your younger daughter seized in the orchard and turned blue, and you ran screaming back to the house with her in your arms, unable to open her jaw.

I check up on you, yes.

I know you have vintage items online. You and your daughter

model them. You've gained weight since the second child. All your girlishness is gone, replaced by a heaviness in your stance and a look of desperation, like you've lost something that can't be replaced.

I mean this gently.

Your home is surrounded by forest, lake water, crabgrass, dirt. You watch the sunset, a brilliant orange, from your porch. Your hair is the longest I've seen it, red and gold. Your skin is freckled suede.

In one photo, you are smiling from the front end of a canoe with two oars in the water. Your body is strong and comfortable. The person holding the camera saw you so well. You think of this moment often. I know you.

In another photo, you are bikini-clad and midsentence in a deck chair between two of your friends. Each of you gestures obscenely toward the camera. The day is hot, yellow. You had fun that day.

I miss you.

I know you married among trees and surrounded by friends. Your daughter is in the awkward prepubescent phase we were in when we met. Her teeth are too big for her head, crooked, clean.

I wonder who your daughter's friends are, if she loves them like we loved each other. I wonder if she's anything like you in the way she loves them. I wonder if they're good to one another.

In another photo, your husband is shirtless, paunchy, his long hair a mop over his shoulders, squinting into the Florida sun with your younger daughter's legs around his neck, holding her by the feet. Behind him, a body of water sits blue and calm under an unclouded sky.

YOU HAUNT ME IN MY EVERYDAY. When I think of you, you are twelve, our most perfect age. You're laughing, and the gap between your front teeth is half the size of mine—we are comparing them. This is the before time: before the real hurt came. We exist in the perfect sweetness of girlhood with our feet in a pool, with matching bathing suits, with egrets stalking through the grass behind us and lizards wending their subtle ways across leaves. We are diving in the water. We're clean.

I open up this time so I can feel all the other time around it. I can see it in sharp focus: a difference of this or that, the light or the dark. I am choosing the light.

In the dark, hurt is pushed to the perimeter and stretched. It is variegated, bold. A bright pink scar.

In the light, I can love you the way I want to. The way you deserve.

I hope you're happy.

Mother-Father God

This immaculate idea, represented first by man and, according to the Revelator, last by woman, will baptize with fire; and the fiery baptism will burn up the chaff of error with the fervent heat of Truth and Love, melting and purifying even the gold of human character.

—Mary Baker Eddy, *Science and Health with Key to the Scriptures*

I reserve the right to change my mind.

—Charles Fillmore

I.

I was baptized in the sanctuary of Unity-Clearwater Church in 1985. A gold winged globe hung above the sanctuary stage. People stayed after the Sunday service for the baptism and sat scattered throughout the old theater seats. I was one month old.

Rev. Leddy Hammock led the blessing. My parents stood near

her on the rust-brown carpet in the sanctuary, two pairs of blue eyes and two sets of chin-length oak-brown hair. They'd been married seven months and sober three years—since joining the New Thought Movement.

Leddy told the story of the disciples asking Jesus to hush the children, and how He'd said that children should not be kept away. For children have a greater understanding of the Kingdom. Unless we become like them, she said, we will not be able to enter His unlimited creative consciousness.

She poured water into a shell and dipped a white silk rose into it. She touched my forehead and chin, crossed my brows, and baptized me in the name of the Father (Divine Mind), the Son (Divine Idea), and the Whole Spirit of God.

She reminded all present that my name means "princess," and that I am the royal child of the most high, endowed with infinite possibilities for good.

Children are a great gift, she said, here to be sheltered and also to teach us.

She gave the rose to my parents. Leddy had known them since they found the church, and she had married them. As they came from different faith backgrounds, they'd held the wedding in an upscale restaurant rather than a place of worship. My mother was three months pregnant. Looking at the photographs, I can see the small rise of myself beneath her sensible white suit.

WHEN I WAS YOUNG MY PARENTS WERE DEEPLY INVOLVED WITH UNITY-CLEARWATER. CHURCH members' names were omnipresent in the conversations we had at home; the songs I sang in the bathtub were those I'd learned in Sunday school. And yet, for the amount of time my parents devoted to the church, my own memories of it are scarce. I was often with babysitters when my parents were there; I had only the vaguest sense of what they did there. My memories of my parents take place between outside commitments—full-time jobs and church meetings. I had limited access to these places.

When I was twelve, we left the church and never returned. I never learned why. Now, nearly twenty years later, I still wonder. I crave better understanding of my parents and Unity-Clearwater. What it meant to them, why they left. I crave understanding of what the church taught me—the unconscious ways in which it still shapes me.

In my conscious mind, the church has been reduced to a foggy set of images floating through my childhood memories: A wooden replica of Noah's ark big enough for a child to climb inside. The koi pond in the courtyard on the way to the meditation chapel. The small window in the door of the Unity-Progressive Council, which opened onto a long hallway of other doors; knowing my parents are behind one of them. A glass gift case with baubles handmade by church members. Palm Sunday, carrying fronds into the sanctuary—the mustiness of it—singing, "This little light of mine, I'm gonna let it shine." The Prayer for Abundance, manifesting prosperity in the lives of all who spoke it—a wicker basket floating down each row—and its ending, "Thank you, Father-Mother God!"

My mother was one of six children, a middle child. She was fifteen when she moved with her family from Paterson, New Jersey, to Winter Park, Florida. It was 1964. Her stepfather's drinking had grown so intolerable that my grandmother had threatened to move to Florida with or without him—he was only invited if he put down the bottle. Sober for the moment, he joined them. They moved into a duplex. My mother's room, such as it was, was partitioned off with a curtain. There were no beaches nearby, but there were plenty of lakes teeming with alligators.

My mother hated Florida. It was hot and boring. It rained every day on her walk home. The girls at school were all snooty, the daughters of citrus barons, wearing Villager clothing while my mother wore hand-me-downs. She missed her friends back home in New Jersey, and her brother who had joined the navy, and her older sister, who'd already married. Soon, her stepfather took up drinking again. In the afternoons, while my grandmother was at work, he'd line the kids up and browbeat them until they cried, then make fun of them for crying. My mom dreamed of escape.

Barbara, a girl at school, had also just moved to Florida, from Pennsylvania—not so far from New Jersey. She hated Florida, too, and she and my mom became close friends, bonded by their mutual appreciation for the shortcomings of their adopted state. When Barb got pregnant and her father threw a butcher knife at her, my mom comforted her. When Barb's boyfriend abandoned her, my mom was there for her. Barb's brother, Bob, was nineteen and he liked my mom. Bob was in the navy, like my uncle Dennis, and was close with his parents—the mark of a good man. After just a few months of dating, when my mom was sixteen, he asked her to marry him. Despite her mother's pleading, that's what she did.

In 1966, my mother dropped out of high school and moved to Norfolk, Virginia, where Bob was stationed. They rented a duplex and painted the walls turquoise. It had a tiny living room and an even tinier kitchen. My mom knew no one. An introverted girl in an unfamiliar place, she grew afraid of leaving the house alone when Bob was gone—and he was gone at sea for weeks at a time. When he was home, they argued. Their fights escalated. One night, he punched her. He apologized. My mom stayed.

After the military, they moved back to Florida and he joined the police academy, then they kicked him out. Then he welded for a while; then he didn't. He convalesced. My mom went to work making children's clothing in a factory for $1.60 an hour. Coming home, her arms were dyed red to the elbows. She'd clean up, make dinner, then visit Bob's parents. Every day. Bob's parents sued people—the city, other motorists, etc.—for a living. Like Bob, they were lazy. And fat. At his fattest, Bob weighed 350 pounds. He loved food—without discernment. As he grew fatter, he grew lazier, and meaner. My mom went to work with black eyes, fat lips. No one at work said a word.

When Bob got going, my mom would run and hide in the school yard behind their house. She once picked up a knife, but put it back down, afraid he'd use it on her. He never found her in the school yard, but sooner or later, she'd have to go home. There were no shelters in the area back then, and she had no friends. She was afraid to call the police—they all knew her, and they knew Bob. He'd hurt the dog. He'd hurt her worse if she stayed away longer. With nowhere else to go, she went back. He'd kill her. She grew certain of it.

One night, curled in a ball as Bob punched her, my mom grew calm. She felt that there was a spiritual being inside her. It was a new feeling. Bob could kill her body, but he couldn't kill *her*; he could physically control her, but he couldn't control her mind. He could make her act the way he wanted her to act, but he could never make her think the way he wanted her to think. This was the turning point.

My mom left Bob in 1976. She left the house, the car, the furniture, the checkbook, her clothes, and the dog. For two weeks, she crashed on her supervisor's couch. Then, because she had no one else to call, she asked Bob to help her find an apartment. He told her he would, but that she wasn't allowed to stay in their house in the meantime—another woman was already living there. My mom would have to stay in a motel.

A few days later, she moved into her first apartment. It took weeks to decide what to put on the walls. After ten years with Bob, she didn't know who she was, had never known. She bought a matching black-and-white-plaid sofa and chair set. She asked her mom to help her buy a used Pontiac, pea green, with the $300 she had to her name. Bob had never drank, so my mom had never drank. Now she drank, exclusively beer. Drinking made her sociable. She made friends who drank with her. She went to bars alone. She felt wild.

One night in 1978, my mom went to a biker bar in South Tampa to hear the Mad Beach Band. She'd dated the lead singer for a few months until he'd broken up with her. She'd taken to showing up at gigs and just kind of . . . staring. That night, she wore a striped cotton beach dress. Her hair down. She'd come alone. On her third beer, the band played her favorite song, and she began to sing along—she knew every line. The guy next to her turned to get her attention: "So, you've heard this band before?" He smiled.

The man would become my father. He had light blue eyes, a scruffy beard, and large wire-framed glasses. He wore a baby blue Jimmy Buffett T-shirt.

My father grew up in Cleveland, where he'd worked for his father since the age of fourteen—after school and long days in the middle of winter and during summer vacations—cleaning and hauling industrial chemical drums. The men in the yard drank Wild Turkey. So did my dad. He learned the motto "Work hard, play hard" from his father.

But my dad felt work in the barrel yard wasted his time and his abilities—it interfered with his outside interests, his social life, his garage band. He respected his father but didn't want him as a boss. He saw in journalism a way to use his mind, to engage with the world around him in a meaningful way. He saw in Florida an escape.

He graduated from Ohio State and wrote for a year at the *Marion Star.* In 1976, he moved to Florida and became a staff writer at the *Tampa Tribune.*

On the night they met, he asked for my mom's phone number, which he wrote on the back of a matchbook and promptly lost. He looked everywhere for her for three weeks before he found her and asked her out again. Five months later, they bought a house.

By 1982, my dad had left the *Tampa Tribune* and was writing for the *St. Petersburg Evening Independent*, covering municipal governments and the police beat—"looking for mistakes that people made." He'd been reporting for seven years. The stories he engaged with on a day-to-day basis had shrunk the scope of his worldview to prostitution sweeps, drug sweeps, sweeps of public restrooms. People who'd fallen on hard times or tragedy. The world was cruel. My dad drank to cope. He smoked pot and dabbled in cocaine and other hard drugs.

My mom drank heavily, as well. One day, she was supposed to meet the man who was installing new carpet in their home after work. She forgot and went to the bar instead, like she did every night. When she came home, my dad was pissed.

"You can't tell me what to do!" she yelled in return. "Who do you think you are?"

Then she realized what she'd said. She recognized those words: she sounded just like her stepfather. My mom was shocked. She had vowed never to be like Doug, ever, in any way. And here she was saying the same things he did, probably doing the same things that he used to do.

She moved out to get sober. My dad slept alone. He got a teddy bear and a roommate. He turned thirty.

My parents tried to be friends. When my mom began a program of living sober, my dad followed. He got a mentor and took his first steps, but when he got to one that asked him to search for a higher power, he panicked. He had been raised culturally Jewish, but had drifted away from the Jewish community. Considering his recent lifestyle, he didn't feel comfortable with God. He had turned away from any attempt to contemplate, let alone make a connection with, a spiritual presence. He didn't even think to call himself an atheist; he just abstained. Now asked to search for a higher power, he floundered. Presenting himself to an anthropomorphized god embarrassed him. He looked for any kind of sign, or spiritual connection that he could buy into. One Sunday, his roommate invited him to Unity-Clearwater, a New Thought church.

Leddy Hammock had taken over as co-minister just the year before. She was a third-generation Unity minister—a welcoming woman with fiery red hair and a voice like a lullaby. Instead of sermons, she gave lessons. Her lessons were only forty-five minutes: my dad liked that. Part of the lesson was meditation. He liked that, too. He liked the positive, reaffirming messages he heard: Leddy said that he was the creator of his own world, that he had the power to make himself happy and make things happen in his life. She taught that "thoughts held in mind reproduce after their kind," and that sins are errors of thought. That people are not punished for their sins but punished *by* their sins. She said that there is no such thing as a punishing God, that God is not a man or woman, that God is spirit and principle.

He began reading all the New Thought authors he could: Emmet Fox, Charles Fillmore, H. Emilie Cady, Thomas Troward. He drove the church bookseller crazy with how quickly he'd finish one book and ask for the next—he began to build a New Thought library of his own. He was fascinated by Ernest Holmes's book

Science of Mind, which teaches that people are at all times engaged in their own transformations. He started listening to meditation tapes by the spiritual author and modern-day mystic Joel Goldsmith, seeking out the intuitive guidance of what Goldsmith called his "still small voice."

He made friends with people at Unity-Clearwater, including Leddy, and invited my mom to meet them, too—she fit right in. My parents got back together. They were still sober and went to church every Sunday, talking for hours about the class they were taking together, studying H. Emilie Cady's *Lessons in Truth*. They got engaged and rented a new house on Allemande Drive with zigzag wallpaper. My mom got pregnant.

The closing decades of the nineteenth century were a confusing time for the American establishment. The Thirteenth Amendment had recently been ratified. Blacks were moving into cities and vying for the same homes and jobs as whites. Increased industrialization was reducing the demand for men's physical labor. More and more, unemployed men found comfort in bars, opium dens, casinos, and brothels, which inspired the rise of moral reform groups largely led by women. The number of women in higher education also increased dramatically, as did the number of women in the workforce. Organizations dedicated to their political advancement sprang up like daisies—Wyoming and Utah had already given them the vote. The closing of the nineteenth century was a time for women.

Meanwhile, research in the scientific and medical fields was raising questions about the unseen world. James Clerk Maxwell's theory of electromagnetic radiation showed that electricity, magnetism, and light were manifestations of the same phenomenon. Edison's invention of the electric bulb illumined every living room. The telegraph circulated information with unprecedented ease. Germ theory, popularized in 1876, introduced the general populace to the theory that sickness was caused by outside forces, not internal weaknesses.

Hypnotism was also popularized, and its effects were widely debated, dominating medical journals between 1882 and 1893. In one school, Jean-Martin Charcot, head of the Paris women's asylum the Salpêtrière, claimed that hypnosis was similar to hysteria and related to internal weakness. Conversely, Hippolyte Bernheim, professor of internal medicine at Nancy-Université, believed that hypnotism operated on the power of suggestion. The debate gave new, scientific validation to more general fears of contagion brought about by germ theory and other studies of the unseen.

This debate over the extent to which the ethereal could influence the material inspired further exploration into the unseen. Many women saw hypnotism as an opportunity to bring about a new era in which mind ruled matter. The financial, political, and social disenfranchisement of women meant their abilities to effect social change had been contingent upon their abilities to influence the thoughts and behaviors of men. But their claims to power had previously been based upon self-sacrifice and spiritual superiority on the grounds of moral purity, reinforcing the association of women with the "feminine heart"—as opposed to the "masculine intellect," or mind. Now studies in hypnosis challenged that, suggesting women's own minds might be better suited than their hearts to empower them—if they could harness the mind's potential. Their minds, not only their hearts, would save them, and save the world.

In 1862, Mary Baker Eddy traveled from Rumney, New Hampshire, to Belfast, Maine, to see a famous healer named Phineas Parkhurst Quimby. Quimby had been conducting experiments with mesmerism since the 1840s, after observing the demonstrations of French mesmerist

Charles Poyen, and had developed a method of hypnotic healing that appealed to Eddy. According to Quimby, a patient's own belief in the effectiveness of a given cure did more to promote her health than the cure itself. Similarly, diseases themselves stemmed from the minds of the sick, who could be cured by redirecting negative thoughts. The reasoning could be done either aloud or silently in the mind of the healer. In the ten years prior to Eddy's visit, Quimby had reportedly cured thousands of people around New England.

Eddy was the youngest of six children born to a quick-tempered and punishing father. Since childhood, she had suffered from a nervous sickness that resulted in fainting episodes, sometimes rendering her unconscious for hours at a time. She treated chronic indigestion with a strict diet of water, vegetables, and bread, at one point eating just once a day. Having found traditional medicine to be ineffective in relieving her symptoms—not to mention riddled with undesirable side effects and risks—Eddy had turned to allopathic medicine and alternative therapies such as homeopathy and hydrotherapy to treat her mysterious illnesses.

Eddy's first husband died suddenly in 1844, when she was six months pregnant. The shock left her bedridden for months, during which time she became interested in mesmerism, animal magnetism, Spiritualism, clairvoyance, and the stories of Jesus as a healer. Her son was sent away to live with relatives, and though Eddy's second husband promised to become the boy's legal guardian, he never followed through, and Eddy lost touch with her son for the next twenty years. Her new husband was unfaithful, and often traveling for his work as an itinerant dentist; he would eventually abandon her. Detached from familial ties, and desperate for any relief from her misery, she decided to visit Quimby, whose reputation had preceded him.

After working with Quimby for a short time, Eddy pronounced herself cured. For the next three years, Quimby tutored Eddy in his brand of mesmerism, and they kept in close touch until his death in 1866. In accordance with his beliefs about the mental origins of sickness, he considered Bible stories and church sermons meant to frighten practitioners into submission a root cause of many ailments. To treat these cases, he would furnish his patients with reinterpretations of Bible stories he'd penned himself, meant to free them from their fears. Taking a cue from Quimby, after his death Eddy penned her own reinterpretation of scripture, and in 1875 published *Science and Health with Key to the Scriptures*, which became the foundational text of her new religion, Christian Science.

The central principles of *Science and Health* are threefold:

First, God is all.

Second, God is good and therefore everything is good, and true evil doesn't exist.

Third, God is Mind—or Divine creative force—and therefore the true world is pure spirit, immaterial. What appears to exist as the material world is an illusion caused by another creative force: "mortal mind," or human thought. Unlike Divine Mind, mortal mind is temporal; it dwells in the illusion of material reality, ruled by the faulty reports of the senses. It believes in evil and, by believing in it, manifests it in the illusionary world.

It follows, then, that as children of Divine Mind, humans have godlike abilities and can overcome the appearances of evil, which they bring upon the world: struggle, sickness, and death. In order to do this, one must repent sins and let go of the unreal—material reality—relying instead on God, or "immortal and omnipotent mind." One does this by silently "arguing" away the appearance of evil.

Christian Science practitioners were trained in healing techniques and claimed to be able to argue away financial difficulty, relationship troubles, ill health, and any other scourge that might befall patients' material perceptions. Healing was deemed more easily forthcoming if the patient was receptive. However, one need not be intentionally receptive to be influenced by others' thoughts. Similar to Quimby's ideology, in Christian Science the evil thoughts of others make us sick. Likewise, believing that said evil has no reality makes us well.

Christian Science spread like wildfire. Soon, Eddy was giving lectures to packed theaters around the northeast and teaching classes to students who delivered the good news far and wide. One of these students was the New Thought "Teacher of Teachers" Emma Curtis Hopkins, then a sickly thirty-two-year-old New Hampshire mother and housewife to an insolvent and abusive husband. In December 1883, Hopkins was healed by a student of Eddy's. Later that month, she journeyed to Boston to study with Eddy herself. With a convert's enthusiasm, by January Hopkins had committed fully to the advancement of Christian Science, writing to Eddy, "I lay my whole life and all my talents, little or great, to this work." That September, Eddy appointed her editor of the fledgling *Christian Science Journal.*

When my mom went to work for the Largo Police Department in Largo, Florida, in 1984, they hadn't had a victim advocate in months. In the two years since joining the New Thought Movement, she'd earned her master's degree in psychology while working full-time as a case manager for Boley, a mental health agency. One afternoon, a client wandered away from his group home and was picked up by police. He was schizophrenic and was so distraught his speech was unintelligible to the officers. All they could make out was my mother's name. They called her and she came immediately. Though the officers hadn't been able to calm the man, he went away with my mother easily. When she later applied for the position of victim advocate, she was hired on the spot.

A stack of old crime reports sat atop her new desk. She began to pick through them. In one, a woman had been raped and left for dead behind a bar. In another, an intruder had raped an elderly woman in front of her husband. There were many cases of child abuse and many of sexual abuse, but the vast majority—75 percent—were domestic violence cases.

That year, Florida legislators toughened the state's domestic violence laws, enabling police to make arrests on misdemeanor assault or battery charges even if they hadn't witnessed an incident—typically a requirement for misdemeanor charges. To do so, the law said, they must have evidence that a victim had been injured or have probable cause to believe that an injury could be inflicted if they didn't make an arrest. But at the time, there was no state- or countywide protocol for responding to domestic violence calls—there wasn't even a domestic violence detective on the Largo police force. Officers would show up, walk the offender around the block, then take him home. Though domestic violence was legally a crime, general thought was that there was no point in arresting an abuser: they'd just have to release him later, and then he'd go and do it again. Often, victims recanted their claims, which seemed like a waste of everyone's time to the officers. It seemed to suggest that the claims weren't true, or were exaggerated at the very least.

Also that year, the Florida legislature passed a law enabling victims to take out injunctions for protection against their abusers—but only if they were married—and even then, there was no mandatory penalty for an abuser violating his injunction. Besides which, law enforcement generally considered matters between spouses to be private; they were a family's business, not the state's, and state attorneys didn't like to prosecute them. If an officer even wrote a report when responding to a call, he most likely soon forgot about it.

By the time my mom arrived, detectives at the Largo Police Department were used to working without a victim advocate. When officers responded to a rape call, or a child sexual abuse call, or a domestic violence call, they were supposed to call my mom to the scene. She was to connect the victims with services or housing, and assist them with safety planning and legal aid throughout the court process. But the officers would forget my mom was there, and when they remembered, they wouldn't use her. Her desk was in the lobby; detectives were in the back. Though this made her available to homeless people who came in off the street looking for housing, it made it hard for my mom to get involved with criminal cases.

My mom battled to get officers and detectives to include her; some of them didn't think the role was necessary and refused to call my mom at all—one officer believed that women were never raped, and that claims of rape meant they were "hiding something" like infidelity or prostitution. No point in calling a victim advocate to the scene if the victim wasn't a victim. Others thought it was a pain in the ass to have to keep her informed as to a case's progress.

Once, at an officers' training session, my mom stood up in front of the class and announced, "I'm your victim advocate! Use me!" They all laughed.

Such was the culture. At Christmas, officers hung their favorite mug shots on the station tree as ornaments. Black humor kept them sane while they were doing their job—caring too much about victims, on the other hand, didn't help. Getting sensitive would get them in trouble. Teasing among them was merciless.

Domestic violence cases continued to flood in. My mom fought to work on them. Through her studies at Unity-Clearwater, she had come to believe that no one was purely evil. People could behave in evil ways, but ultimately they were all children of Spirit, like her, and thus on the path to spiritual enlightenment. It was her role as a student of Truth to make them realize their potential. With some it was harder than others, but if she persisted, love would prevail, so she pushed on.

Then one day, her desk was gone from the lobby. Someone had moved it to the back among the detectives—one of them in the Crimes Against Persons division saw the value in a victim advocate and began encouraging others to give my mom cases. Soon, they were calling her out daily on cases of child abuse, sexual abuse, rape, and attempted murder. But never for domestic violence. If she went on those calls, they said, she'd be out every night. So they left the reports on her desk. It was an improvement.

That year, the local chapter of the National Organization for Women (NOW) began lobbying the Police Standards Council for a countywide policy for police to arrest abusers every time they responded to domestic violence calls, and track domestic violence arrest and prosecution rates. They wrote letters and made phone calls and attended county commission meetings—and made themselves general pains in the ass. It took two years, but in 1986, the county passed the policy.

Soon after, the Standards Council formed a Domestic Violence Task Force, which included the director of St. Petersburg's domestic violence shelter, Community Action Stops Abuse (CASA); a leader from NOW; and other advocates and government leaders, including my mom. She couldn't change what had happened to her, but she might be able to stop it from happening to others. And where she couldn't prevent it, she could change the way people were cared for afterward. It was a matter of changing people's perceptions—inside the government, but the general public's as well.

As editor of the *Christian Science Journal,* Hopkins had enlarged it and made it a monthly publication. She also made a point of skewering anyone whom Eddy perceived as having violated or misappropriated her teachings. This included Eddy's own students: in the decade before meeting Hopkins, Eddy had taken several to court for using her work. She had returned to a puritanical religious approach in Christian Science, eschewing all other schools of metaphysics—including the Spiritualism, clairvoyance, and other mystical avenues she had once dabbled in herself—especially those that taught variations of Christian Science. She considered herself something of a prophet with a sole claim to truth; those who strayed from the truth, or misrepresented it, were frauds and heretics.

Eddy was also sensitive to possible rivals. In the year since her editorship began, Hopkins had shown herself to be an important actor in the Christian Science movement. Loyal to Eddy, Hopkins was also ambitious and charismatic, with a decidedly countercultural streak and a taste for religious eclecticism, reading voraciously from all faith traditions. In 1885, thirteen months after hiring Hopkins, Eddy abruptly dismissed her.

In the months prior to her termination, Hopkins had befriended another energetic and enterprising former student of Eddy's, Mary Plunkett. With encouragement from Plunkett, in late 1885 Hopkins left her husband and son to move to Chicago, where the "mind cure" movement was booming. By spring 1886, she'd founded the Emma Curtis Hopkins College of Christian Science, naming Plunkett as president. Shortly thereafter, she founded, alongside a group of prominent students, the Hopkins Metaphysical Association, an open forum for people interested in metaphysical healing of all kinds. Members of the group went on to establish independent branches, and, by the end of the following year, between seventeen and twenty-one Hopkins Metaphysical Associations were operating across the country. Hopkins toured and gave lectures to audiences sometimes numbering in the hundreds. By the end of 1887, she had personally instructed six hundred students.

Like Eddy, Hopkins taught that God is all, God is good, and God is Mind. Therefore, matter cannot exist. Therefore, there is no true evil. Therefore, health and spiritual redemption can only be attained once God's true nature is understood.

Though she called herself a Christian Scientist, unlike Eddy, Hopkins strongly opposed teachings about sin and repentance, claiming that such teachings created evil by implanting false beliefs. Instead of instructing her students to "argue" evil away, Hopkins taught them to "enter the silence" and meditate on "affirmations" and "denials" to cleanse the mortal mind of false beliefs and enable the Divine Mind to shine through it. In addition, practicing such affirmations and denials empowered believers to embrace their Divine nature and become creators of their worlds. In her tract "The Radiant I AM: A Self-Healing," filled entirely with such affirmations and denials, Hopkins demonstrates: "I AM power of Life to the universe. Because I live, all that hath form or name shall live. There shall be no death nor fear of death throughout the boundaries of eternal spaces from this day forth forever. That which proceedeth forth from Me is Life and the power of Life forever." Significant among Hopkins's teachings is her conceptualization of the Trinity. Eddy had believed the Trinity was suggestive of polytheism, so was against it, but spoke of God as being both Father and Mother—a masculine, or material, aspect countered by a feminine, or spiritual, aspect—with the two constantly at war. In Hopkins's ideology, God is Father, Son, and Mother-Spirit, or

"Holy Comforter." Each corresponds to a different historical epoch: God the Father represents the patriarchy of the past; God the Son represents the Second Coming of Christ and the freeing of human thought; and God the Mother-Spirit represents the dawn of the new era. Together, they are the whole spirit of God.

My mom became director of the Spouse Abuse Shelter of Religious Community Services in 1986, the year Pinellas County passed its preferred arrest policy. I was one year old. The shelter was a dreary two-story house with few windows. It had a big, drafty living room and bedrooms off the living room with enough space to sleep twenty people. The shelter had built a kitchen and an area upstairs for the house manager. On Christmas, a man in a Santa Claus costume passed out candy to any children staying there.

At the time, there were six and a half people on staff, including my mom: a secretary, two counselors, an overnight house manager, a day care worker, a part-time house manager for the weekends, and occasionally a volunteer. As director, my mom supervised the staff, wrote grant proposals, spoke at public events—even answered the phone. It was a heavy workload. Stress made her irritable all the time. Now four years into her sobriety, she'd learned not to lash out; instead she stuffed her feelings down. She carried her anger around like an anvil, pretending it wasn't heavy, lest she feel its full weight and let it fall.

Though the preferred arrest policy had been instated, police response continued much as it had before, and there was still no penalty for abusers violating injunctions for protection—they could rip up their injunction in front of a police officer and the officer couldn't do anything about it. There were no domestic violence units in state attorneys' offices, and no protocol for prosecuting abusers if victims refused to testify. In those instances, the cases were dropped. My mom wanted all of these things to change—and more. She knew what it was like to feel unsafe in your own home; it was an unacceptable way to live. She wanted a cultural revolution. She wanted to bring about a nonviolent world.

In addition to her work with the Domestic Violence Task Force, and the trainings she was doing for local police—who misbehaved in class and often mistook her discourse on male privilege to be evidence of lesbianism—she was also a member of the Florida Coalition Against Domestic Violence, as was every shelter director in the state. Every few weeks, she'd leave at four in the morning to drive to Tallahassee for meetings, and to lobby legislators for better laws. The coalition wanted a statewide arrest policy like the one in Pinellas, stronger injunctions, mandatory training for state attorneys, and laws to protect police officers from legal action if they chose to arrest abusers. They also wanted to expand the legal definition of domestic violence, to require judges to set higher bail for abusers, and to prevent abusers from being able to buy guns. If they couldn't get a meeting with a legislator, they showed up in the legislator's office. If they couldn't get a meeting that day, they'd wait until the next day. They worked with legislators' staffs to draft language for bills, and testified at committee meetings, and marched on Tallahassee for abortion rights, and drove to DC to march for the Equal Rights Amendment.

When she wasn't lobbying legislators, and training, and marching, my mom was hosting NOW meetings in our living room, and meeting once a month with the task force—and outside of that, she was taking classes at the church, and raising a child, and spending time with my father. Often it still didn't feel like enough.

Women at the Spouse Abuse Shelter could stay for thirty days. Back then, there was no transitional housing, so women who came through the shelter and couldn't find other living arrangements were forced to return to their abusers. There was nothing my mom could do for those women but keep the doors open. One morning, walking through her secretary's office, my mom passed a woman whose body was covered in bruises. Her eyes were flat. She was one of many like her, but something about her struck my mom. She proceeded past the woman into her office, shut the door, and burst into tears. She'd never understand how someone could do that to another person. She couldn't tolerate the idea that this woman might have to return to an unsafe home. My mom sought solace in her faith, went to church every Sunday, read books, prayed often, but where was solace for these women? She was burning out.

In 1888 Hopkins and Plunkett parted ways, and Hopkins assumed the presidency of the College of Christian Science and the Hopkins Metaphysical Association. The preceding months had been trying for Hopkins; differences between her and Plunkett had come to a head, and Plunkett had absconded to New York with the mailing list for *Truth.* The journal was the organ of the college and connected the school to Hopkins's outreach satellites nationwide. As a result of this breach in communication, many of those satellites seceded or aligned with Plunkett, who was now their main contact in the New Thought Movement. It was time for Hopkins to take radical action.

That same year, Hopkins transformed the College of Christian Science into the Christian Science Theological Seminary, deemphasizing the professional aspect of Christian Science and emphasizing ministry. Believing she was on a spiritual mission, she met one-on-one with every advanced student in his or her final year, developing an individualized curriculum for each.

The seminary's first ordination ceremony was held in January 1889. Suffragist Louise Southworth was guest speaker. The class was comprised of twenty women and two men, marking the beginning of a new spiritual epoch for women in divinity. In her speech, Southworth proclaimed: "Divine Truth has come at last to give woman her proper status in the world." In the seminary's lifetime, Hopkins would ordain hundreds of female ministers. Many would go on to become prominent members of the New Thought Movement, or to found churches of their own. One of these churches was Unity.

II.

In 1886 a young, bookish housewife named Myrtle Fillmore attended a mental-healing seminar in Kansas City, Missouri, held by the New Thought teacher E. B. Weeks, a student of Hopkins. Myrtle had contracted tuberculosis at a young age and tried myriad courses of traditional medicine to cure it. When the Kansas City real estate market crashed, sending the Fillmores into debt, Myrtle's symptoms flared and her doctors recommended she leave Kansas City for preferable climes. Desperate for relief, she signed up for Weeks's seminar, where she heard the teacher utter an affirmation that would change the course of her life: "I am a child of God and therefore I do not inherit sickness." She abandoned traditional medicine and devoted herself to faith, attending every metaphysical lecture that came through Kansas City.

By 1888, Myrtle considered herself healed, attributing her cure to prayer alone. The following year, she and her husband, Charles, began publishing their own New Thought journal, *Modern Thought*, later called *Unity*, and in 1891, Hopkins ordained them at the Christian Science Theological Seminary. It was the heyday of New Thought, and churches were springing up in diffuse locations across the country, each with a slightly different approach to the power-of-mind philosophy inspired by different leaders of the movement. In the tradition of Hopkins's thinking, the doctrine of New Thought did not to adhere to a singular dogma, but rather sought Truth continuously and omnivorously.

Though Myrtle supplied the original impetus for founding what was then called the Unity School of Practical Christianity and was an active participant, Charles became the principal leader of the Unity movement and its most active scribe. Despite a lack of formal education, since childhood he had immersed himself in Shakespeare, Tennyson, Lowell, and Emerson, and works of Spiritualism, the occult, and Eastern religion, and he was a talented writer. Charles had rarely attended church as a child but, like Myrtle, was deeply spiritual and eclectic in his theology. Despite his esoteric tendencies, his approach to metaphysical Christianity was decidedly practical—indeed, Unity is often referred to as "Practical Christianity."

Like Hopkins, the Fillmores taught that God is all, God is good, and God is Mind. Affirmation and denial are cornerstones of the faith, as is the application of such statements in healing. Practitioners are taught to affirm health, well-being, wealth, and safety. These affirmations at times ring so practical as to border on humorous: some early pamphlets and articles published by Unity feature titles such as "Curing Colds," "An Airplane Blessing," "An Automobile Blessing," and "A Salesman's Prayer."

Also like Hopkins, the Fillmores regarded the Bible as both a sacred text and a piece of literature with a complex history. Their interpretation of it was strictly allegorical. Unity distinguishes between the Jesus of history and the divine Christ, emphasizing the "Christ consciousness" and "Christ in you" that each person possesses. While Jesus was a master teacher, says Unity, *Christ* is the divinity of all people; while Jesus was Divinity in human form, he was not special in that sense, as we are all Divine. We can all do what Jesus did.

At the insistence of Mary Baker Eddy, the Fillmores abandoned the name Christian Science and adopted that of the Unity School of Christianity. The choice of new name is telling; it was the Fillmores' intention to establish not a church, but an educational movement. It was only for the purpose of teaching others how to transform their lives with Practical Christianity that Charles finally sought to make clear to himself and others what he really thought and believed. Even then, he did so in a set of twelve informal lessons based on his own personal experiences, which he taught in an informal, discussion-based format over two weeks. Under pressure from students seeking teaching certification or ministerial ordination, Unity finally came to be formalized as a church.

This presented a challenge. The New Thought Movement had stressed individual reliance, using one's own inner truth as a guide. Now people were being taught the principles of Unity from a definite point of view, and going on to establish churches of their own bearing the Unity name. Charles had always resented Eddy's centralized doctrine. It forced the Fillmores, having diverged from it, to stop using the name Christian Science. Nonetheless, a Unity field department was established and Unity ministers naturally organized themselves into a ministers association. They set standards for groups that wanted to be Unity Centers and set the qualifications for students intending to minister and teach.

The standards were strict; the organization's bylaws said that any group that wished to be recognized had to adhere to the Unity teachings and textbooks—and get rid of any texts and teachings that didn't conform to the Christ Standard as recognized by the

Unity School of Christianity. Though the seminary curriculum was meager compared with the standards of other churches, Unity was still a young religion. It was expected that the standards would be raised with the passage of time.

In May 1989, one month before I turned four years old, my father ascribed his name "in prayerful openness to the Spirit of Truth and with the guidance of God" alongside three others', founding the Unity-Progressive Council (U-P.C.). Never one to hover around the fringes of things, in the seven years since he'd joined Unity-Clearwater he'd become a vital member of its inner circle. Concern had arisen among them that the Unity religion was becoming diluted by churches whose practices were not in line with its foundational teachings. In recent years the lines differentiating Unity from other religions, New Thought or otherwise—and even from some New Age belief systems—had begun to blur in ways Unity-Clearwater's senior members found disquieting. The U-P.C. sought a return to fundamentals.

At the first U-P.C. meeting, Leddy Hammock sat at a six-foot table in the future U-P.C. headquarters at Unity-Clearwater with the other three founding members, including my father. As she listened to the group discuss their plans, she decided that even if she had to give up everything else she'd ever dreamt about in her life to be at this meeting, she would still be there by Divine appointment. The U-P.C. was that important.

After they had signed their names to the Progressive Reaffirmation of Unity Faith, the statement of faith of their new Unity movement, the meeting adjourned. They stepped out into the night. Leddy and my dad looked up at the stars. They'd involved themselves in a heroic endeavor, setting aside everything else that they cared about: families, professions, homes. In returning to the fundamentals of the Unity faith, they were restoring Truth to the tradition and teaching others its power. They had chosen this. It had chosen them.

"Friends. That's what it's all about," said Leddy. "We have such friends in this. Most people never have."

"Yes," said my dad. "And we always will."

So it was, and so they let it be.

In 1948, Charles Fillmore "made transition"—New Thought terminology for death—and his son Lowell assumed presidency of what was by then called Unity School. It was a decisive event in Unity's history: Lowell was a religious moderate and had a vision of growing Unity's reach in America. For the first six decades of Unity's life, Charles Fillmore had been its sole director, ensuring that the identity of Unity was consistent with his vision, but all that was about to change. Lowell's strategy was to try to position the Unity movement closer to the religious mainstream by allowing Unity churches to become autonomous and by promoting practices like crystal healing and channeling, or conversing with spirits—practices Charles Fillmore had specifically spoken against. The Unity movement continued to grow.

After Lowell's transition in 1972, his nephew Charles Rickert Fillmore took over. Then Charles Rickert passed and his daughter Connie Fillmore Bazzy assumed his place in 1987. Control of the mission was handed off so many times that the path of the Unity religion swerved far outside the route Charles Fillmore had originally planned for it.

Elsewhere, a new institution had arisen: the Association of Unity Churches, or AUC. Whereas Unity School had been a center for prayer, publication, and ministerial training, though not accredited, the AUC had come about in response to a perceived crisis in doctrine and a failure of leadership on the part of Unity School, which had come to have little to do with the operation of Unity churches. Now the AUC served as the organizing body of as many as five hundred Unity ministers. Lowell's revisionist leadership had caused a de-emphasis in the movement on such central concepts as regeneration, or reincarnation; the Father-Mother God construct; human divinity; and the pitfalls of the traditional Christianity institution described in Unity's literature and teachings. Whereas there was intellectual commerce between the AUC and Unity School, they were functionally and philosophically independent. This gulf between the two leading Unity institutions resulted in a contradiction, and resultant vagueness, of self-definition that frustrated members of the U-P.C. when they were asked fundamental questions about their religion.

Unity was by this time a hundred years old. Its prayer center, Silent Unity, was receiving over a million requests for prayer annually. In the first of three courses required for members of the Unity church to become Participating Members of the U-P.C., Dell deChant, one of the U-P.C.'s founding members and associate pastor at Unity-Clearwater, lists some typical difficult-to-answer questions facing a Unity practitioner in 1989:

> What is Unity? Is it Christian? Is it part of the New Age movement? Is it part of New Thought? Does it endorse channeling and channeled teachings? Does it find any unique or authoritative value in the teachings of the Fillmores and H. Emilie Cady? Where can someone find out what Unity believes? Is Unity consistent in its teachings? What are its teachings on the central elements of Christian doctrine? These are religious questions, and there is no coherent, unified, and authoritative answer to these questions given by the two major institutions that represent the Unity movement.

This vagueness seemed to extend into Unity's educational offerings, the U-P.C. asserted. Graduates from the Unity Ministerial School and the Unity School of Religious Studies weren't well versed in Unity's teachings or its history, or theology in general. Thus, they weren't good ministers, let alone representatives of Unity. In fact, Unity didn't even have an accredited theological seminary at the time—no New Thought church did. The U-P.C. also complained that students in Unity schools were being taught by teachers who lacked terminal academic degrees, and they followed a curriculum that was less academically rigorous than that at other seminaries. When the founding members of the U-P.C. had attempted to reach out to Unity headquarters in Unity Village, Missouri, with their concerns, they were rebuffed. They were left with no choice but to organize independently.

AS A FOUNDING MEMBER OF THE COUNCIL, MY DAD'S CHIEF CONCERN WAS THAT THOSE IN need of Practical Christianity might not be able to receive its benefits should it continue to be adulterated by what he saw as untrue teachings. He had seen the benefits in his own life and didn't want others to be denied: People who were sick or battling addiction might not be able to heal themselves. People who were struggling financially might not realize that they had the power to manifest wealth in their lives. He saw the need to address the challenges of identity, authenticity, and institutional structure in Unity.

The U-P.C. proceeded on the basis of three affirmations:

1. To reaffirm the traditional teachings of Unity.
2. To establish a democratic and egalitarian Unity institution, and encourage the establishment of similar institutions.

3. To expand and enrich educational opportunities for practitioners of Unity.

The U-P.C. launched the Unity-Progressive Press and began publishing courses, tracts, and pamphlets, including one entitled "A Progressive Reaffirmation of Unity Faith." By this time, my father had left journalism and gone into advertising, so he designed and printed it, as he did all of U-P.C.'s publications. The pamphlet is cream-colored and trifolded on letter-sized paper, and features on its cover a cheerful cartoon winged globe, a symbol of the relationship between Spirit, soul, and body in the Unity tradition. It was inspired by the cover of the Fillmores' *Unity*, in which the thirty-two-item Unity "Statement of Faith" was originally printed. Most of the material contained in the new one is also in the original. The U-P.C. pamphlet lists forty statements expressing Unity's primary teachings. Those wishing to be members of the U-P.C. are asked to sign their names at the bottom.

Education was the most participatory aspect of the U-P.C. Among the council's initiatives was establishing "a school of higher learning" in the form of the Unity-Progressive Theological Seminary. There, excellence would be encouraged and students could study not only New Thought teachings but also Christian history and theology, and ultimately receive ordination. Those seeking teaching certification but not ordination could proceed from the Participating Membership courses on to the Certified Instructor Program, a fourteen-course program with at least seven classes in each, taught over two years. My mom wasted no time in signing up for the latter.

As part of her U-P.C. coursework, my mom was required to keep a prayer journal. She kept it in the same spiral-bound notebook she used for a counseling group she'd joined after getting sober. The entries are brief; she writes a few lines every day, maxing out at half a page. When she begins the journal, she's enthusiastic about recommitting herself to the daily practice of prayer after several weeks of not praying or meditating consistently—"I have had a challenge with discipline," she admits. She reads from Frances Foulks's *Effectual Prayer* each night and calls Foulks's writing beautiful, her openness to Spirit so complete. She affirms the prayers at the end of each chapter, sometimes reciting them over to herself several times, other times just holding them in mind. She's especially moved by the chapter on forgiveness. She finds herself wanting to remain in it even as she moves on to the next chapter. She thinks of the teachings in relation to a colleague with whom she's in conflict: "I was able to thank the person I had been focused on last night for the things she had given me: self-awareness, a clearer sense of my role in my job," she writes. In her first weekly self-assessment, she already finds herself stopping during the day and "turning more readily to prayer instead of dwelling in some negative consciousness."

After the first week, though, she hears some bad news at the shelter. She spends most of a day dwelling in negativity. The next day is the same: She finds it necessary while driving to pull over and let her mind "chill out" before pulling back onto the road. When she tries to meditate late in the evening, she finds she's too sleepy. The next day she notes, "Did not formally pray," but returns the following day with determination, praying before work and reading from *Effectual Prayer*, which inspires her to turn her mind again toward forgiveness.

The following week, she's frustrated by her failure to be moved by her readings and the lack of time for formal prayer within her busy work schedule. She is nonetheless, in her weekly assessment, "amazed to realize how angry I have been and how dramatically different that feels right now." She sees this as a result of committing herself to the attempt of daily meditation, if not always achieving the act itself. She resolves to be more consistent.

For the next two days, she's in "crisis mode" in Tallahassee, where the Domestic Violence Task Force is lobbying. "I don't recall even thinking about taking the time to pray," she writes later. The following day, she "turns it over" to God and decides to focus in on "naming recent upsetting incidents good" and "being open to learning from them." She's impressed with her ability not to become consumed by negativity like others around her.

She gives a speech for the Catholic Altar Guild the next evening, on behalf of the Spouse Abuse Shelter. Inspired by references to the rosary and the Virgin Mary among those present, she finally, frustrated with her inability to meditate consistently, decides to try reciting the Hail Mary several times slowly, getting in touch with Mary as a "receptive, responsive, obedient state of mind." The familiarity of the prayer—for the first nine years of her life, she attended Mass—and the repetition of it help her break through to a state of peace. "We're all really practicing the same religion," she realizes.

Near the end of the journal, my mom writes about Bob.

> Started prayer time at 10:00 pm, thinking of prayer class and an experience I had many years ago of feeling the presence of Spirit within. I was in a life-threatening situation then so the feeling/memory has not always been all-pleasant. It was tonight. I emotionally felt the Presence, and just let it happen. Joy!

I was twelve when my mom first told me the story of Bob and the Spirit. I'd been having trouble with a friend at school. She was jealous when I sat with other people at lunchtime. She said cruel things about them not liking me and not wanting me to sit there. Her words made the world feel darker. On a particularly bad day, I told my mom about it. "It's like the walls are closing in," I said.

We were in the kitchen. For some reason, the lights were off. My dad had yet to come home from work. I sat on a stool at the island counter and my mom stood on the other side, facing me. I could tell by the way she looked at me that what she was about to say would change the way I understood her. I knew about Bob, but now she told me how he would hit her sometimes just for being there. "Then I realized there's a place inside me that he could never touch," she said. I said nothing, but reached for that place within myself. It looked like a small light floating behind my breastbone. Though I wasn't yet ready to separate from this friend, I felt stronger.

She ends the journal on a Sunday with a reflection on Leddy's lesson of the morning. The topic had been obedience, and it throws my mom into a state of mind she finds hard to explain. Her prayerful experiences are "all being taken in." She compares it to H. Emilie Cady's concept of "chemicalization," or the rapid reconfiguration of spiritual understanding.

> The more I pray regularly, the more it feels my mind (my subconscious) is open but processing so fast that I'm not even sure what is going on. I've learned to name that turmoil good because I always end up stronger and clearer-minded when the dust settles, but it's most disturbing while it's going on.

My mom is an emotionally reserved person. We're close, but she's never been the kind of mother who leaves notes in my lunchbox—though once I asked her to, and she was happy to oblige. When I argue with my husband, I call her for advice. She's supported me in every life choice except the unhealthy ones. In those cases, she saved me. We're alike in our love of silence.

I've seen her cry exactly twice: once when I walked into her bedroom without knocking after her little brother died, the other when I was hospitalized with anorexia. In most situations you can only deduce she's upset about something by studying tiny fluctuations in her facial expression. If you ask her why she's upset, she'll just give you *the look*: a patient glare. I've always mythologized my mom as a kind of warrior, a kind of epic hero. Reading her prayer journal has been emotional. In it, she's human.

My mom finished her Unity studies in 1992, a few months after the Florida Legislature ordered all twenty state attorneys' offices to establish domestic violence units and specially train prosecutors in those units to handle abuse cases. It was a glimmer of hope, and another followed on its heels when the Florida Supreme Court ordered district courts to institute family law divisions, ensuring that judges in such divisions had the aptitude and desire to concentrate their efforts on domestic violence.

Despite the momentum they were gaining, my mom was exhausted. While she adhered to the belief that Spirit was the only true reality, she had neglected the material reality that her physical body had limitations. That year, she resigned from the Spouse Abuse Shelter and the Domestic Violence Task Force, and took a position as the director of a crisis hotline. She devoted much of her time outside of work to the newest phase of the Unity-Progressive Council: the Emma Curtis Hopkins College.

Dell invited my mom to serve on the provisional board of directors. The plan was to develop a new division of the Unity-Progressive Theological Seminary into an entirely independent educational institution—a liberal arts college, in honor of Emma

Curtis Hopkins. It would be a way to raise the level of discourse among Unity's practitioners and its clergy. It would bring serious intellectual consideration back into the Unity tradition. And it would raise the public profile of the religion.

The first meeting convened on June 30 at Leddy's home, across the street from Unity-Clearwater. A small group of PhDs, ThDs, JDs, MAs, and other distinguished fellows gathered in Leddy's living room to discuss first steps. They would offer both bachelor's and master's degrees. Dell had already begun the process of laying out degree requirements and a course catalog, and preparing the articles of incorporation. It was decided that my mom would be president of the board and chair the steering committee—nominations she was reluctant to accept, considering the amount of responsibility they might entail. Ultimately, she did so because Dell, a trusted friend and senior church member, was asking. Among the topics discussed were the necessity of finding funding and the designation of responsibility for beginning work on the thirty-thousand-volume library required of all universities by the Florida Department of Education. They would also need to locate a facility in which to house the college. At the closing of the meeting, my mom led the group in prayer.

Less than a year later, in April 1993, the Emma Curtis Hopkins College board met late in the evening at the home of one of its members to discuss alternatives for proceeding with the opening of the school. They had not been able to raise the target $50,000 set by the steering committee or secure a location, nor had they made much progress on the library. With my mom still presiding as president they decided to raise $30,000 by the next meeting, with the understanding that another $10,000 to $20,000 would be needed prior to the application for incorporation. It was also decided to enlarge the size of the board.

In July, they reconvened at my home. I was eight. I have no memory of this meeting, but I can imagine it easily: our one-story white house with pink trim and a pink rain tree in the front yard, a grapefruit tree in the back. Late evening. Twelve bodies packed into the tiny space shared by our living and dining rooms, their shoes pressed into the stained beige carpet, sounds of my Reader Rabbit computer game coming from my bedroom down the hall. Backs leaning against the white-painted bookshelves built into the outer wall of the kitchen. In the kitchen, a pot of coffee sat on the round wooden table, surrounded by mugs. My mom had cut her hair short by this time, with feathered bangs. She would have been sitting on the floor in her light-wash jeans and socks.

The meeting opened with a prayer led by Dell deChant.

Of the $30,000 they'd determined to raise at the last meeting, just $12,000 had been pledged and $3,000 collected. Due to the shortfall, Dell suggested postponing the application for incorporation, restructuring the board to add members, and aiming instead for an August 1995 opening—two full years away.

Much discussion was had among those present about how to raise funds. Someone suggested reaching out to other New Thought groups in the area. Someone else proposed starting an open-forum series like the one on ethics at University of Southern Florida. The question of temporarily housing the college at Unity-Clearwater was proposed as a money-saving option. Brent Elrod motioned that the board proceed with a 1995 opening as a new goal. Leddy seconded. The motion passed. Leddy closed with a prayer.

That night, my parents read me to sleep. This I remember. My dad taught me to read when I was just three, helping me sound out the opening lines of *The Hobbit*. After that, we'd take turns reading books together, returning to the same stories over and over: *The Velveteen Rabbit*; *The Little Engine That Could*; and my mother's favorite, *Alexander and the Terrible, Horrible, No Good, Very Bad Day*. I was almost too old to be read to by this time—I was reading novels on my own, lots of them—but it was an important ritual, one that we cherished.

When we'd read three books, they left. My door was cracked and I could hear them in the kitchen, talking about the meeting. The U-P.C. had come under criticism from the AUC, which believed they were founding a new religious sect. I fell asleep listening to the murmurs of their concern, and though I was almost too old for this, too, I hugged my teddy bear, the same one my father had bought when my mom left him in 1981, just before they found Unity.

It was around this time when I first felt God was real. My parents had bought me an illustrated children's Bible. I never read the stories in the Bible, just stared at the pictures, which seemed to tell more interesting stories than the ones I found in the text. There were illustrations of men in flowing garments leading groups of others through harsh terrain, a baby among animals, fire, people being healed. The Bible stories meant nothing to me—I don't think I ever learned them, even in Sunday school, where we mostly sang inspirational songs and did arts and crafts. But I knew what I should feel when I read them because I'd felt hints of it before—a sudden thrilling clarity. I could never make that clarity stay; as soon as I sensed it, it dissipated.

Then, one night, sitting in the backseat of my mom's green Oldsmobile Delta coming back from a babysitter's, held the illustrated children's Bible on my lap unopened. It was past my bedtime and I was sleepy. As I looked out the window up into the dark sky, wondering where God was, if he was real like my parents said, I felt a light open up in my chest and spread outward through my arms, my throat, and into my mouth. I watched the dark streets outside my window fly past, knowing, for the first time, that we were made of the same material as all of it, were manifestations of the same omnipotent consciousness. I wanted to speak, to tell my mom what I was feeling, but no words came. Just a feeling of protection. A pure, nameless joy.

My mom has kept a binder of materials from the Emma Curtis Hopkins College on a bookshelf of New Thought texts in my parents' home office for twenty-five years. It's from those materials that I tell this story. In the binder, on the back of a drafted letter from Dell deChant to a potential board member in June 1994 are two pages of another letter, dated the same month and addressed to Alan Rowbotham, president of the AUC at the time. I don't know who wrote the letter—it wasn't Dell, as he's referred to by name in it. I have pages three and six.

From them, I gather that in April 1993, the AUC, of which Unity-Clearwater was a member, conducted a closed investigation of Unity-Clearwater, of which the church was not made aware until later. In response to the perceived actions of the U-P.C. and the Unity-Progressive Theological Seminary, the AUC brought twenty charges against Unity-Clearwater, rendering them "not in good standing" in the eyes of the organization.

That October, before Unity-Clearwater responded to the charges, the AUC sent a letter to twenty-two other churches around Florida sharing "a hostile and negative series of announcements" about the U-P.C. and its alleged policy violations. Some of those churches made the letter public. Word of it reached Unity-Clearwater. The congregation went abuzz with concern.

In response the U-P.C. invited Rowbotham and other AUC members via the letter to the next full council meeting in May 1995 for "dialogue and 'respect building.'" The writer says, "As we affirmed from the beginning, this matter was not handled properly by the Association, and the challenges that have evolved in connection with it have not been caused by us." According to the U-P.C., Rowbotham needed to rescind his negative statements about the U-P.C. until a proper investigation could be conducted following the AUC's own procedure as set out in its bylaws. "This is not a personal matter to us," the writer says, insinuating that it's instead a matter of principles. The controversy only served to stoke the flames of purpose in U-P.C. meetings. The spirit remained one of revolution.

That month, the Emma Curtis Hopkins College board, of which my mother was still president, decided to move forward with incorporating the school and preparing its application materials for the Florida Department of Education. They requested any gifts, financial or otherwise, that the U-P.C. higher education subcommittee desired to share, especially books. By that time, the board had collected just $8,000.

Three weeks later, Bob and Campbell Whitaker, president and dean of the school, came to the board with grave news. In light of the school's lack of finances, a complete library, and a facility—three standards that must be met in the eyes of the state in order to obtain licensure—they believed that denial of the school's temporary license application was a distinct possibility and feared that rejection of this kind would end the project entirely. To guard against this, they advised either further delaying the submission of the school's application or seeking authorization to operate as a religious institution instead of an educational one. The board decided to pursue the latter. In March 1995, they received state approval and the Emma Curtis Hopkins College opened.

In the church directory from that year, I appear in one photo. The tilt of my head is such that I almost didn't recognize myself: I'm all the way to the left in a line of youths on an ice-skating rink with six chaperones. I don't recognize anyone else in the photo, and they are not identified in the directory, but I vaguely remember this day: it was my first time ice-skating. My skates look oversized, as does my sweatshirt—I'm in the awkward middle stage of a growth spurt. I have bangs and my hair is pulled back in a French braid, my favorite style at the time because my mom had to do it for me. My stance is gawky, as if I'm not sure whether or not I'm meant to be there. I'm not sure if these other people know who I am.

My parents are what you might call lenient. I was rarely grounded as a child; instead, we would have conversations about my decision making—how I should know better than to make prank calls in the middle of the night, or why I needed to do my math homework even though I hated math. They went vegetarian when I was four—the Fillmores were vegetarian—but I continued eating meat for six more years. I was allowed to read any books I wanted to read. I could watch almost any movie—only excessive violence was banned. We always debated ideas respectfully. I was taught early that ad hominem attacks were off-limits and that listening to an opponent is the best way to win an argument. My parents had always told me that attending church was my choice. By the time I turned ten, I'd stopped going consistently, riding along with my parents every other week or so.

That same year, I switched to a new elementary school with an arts program. I was interested in dance and couldn't wait to begin; I'd been doing gymnastics for four years and had a feeling that I'd be good at it. I had a hard time making friends, though. I was a shy kid. I spent a lot of time at babysitters'—my parents were busy people and I didn't have siblings who could supervise me in their absence. I had a stutter, which was further isolating, and I'd become well acquainted with loneliness, almost to the point of not feeling it. I turned inward. Many times that year, I lay in bed in the morning affirming that I was sick so that I wouldn't have to go to school. It worked: I missed a third of the school year. By the end of the year, one of my lymph nodes became infected and I had to have it removed. To this day I'm certain I did this to myself through the force of my own mind.

I remember knowing I had a special power. I don't remember being taught this in church—not explicitly. I remember the focus on language, on affirmations and denials. I don't remember the lessons, despite having absorbed them in some deep way. I find this mystery fascinating, how much I am still shaped by something I no longer wholly remember.

What I'm left with instead are snapshots of memories. In the spirit of supporting students' unique spiritual journeys, Unity-Clearwater accepted people from all faith backgrounds: I remember a Passover Seder my father hosted in the fellowship hall, searching for the *afikomen* and finding it hidden under a folded retracting wall—my feeling of pride in that moment. I remember my Sunday dresses: one burgundy velvet, another flower-printed with lace. And hiding in the closet of the Sunday school, eating candy sprinkles meant for arts and crafts. Sitting in a circle singing hymns with other children. And sitting in the dim, Christmas-decorated sanctuary between my parents, holding a white candle ringed with white paper. Knowing we'd reached the end of the Sunday lesson when we stood for the Prayer for Protection. Even now the words bring tears to my eyes. I still feel safer when I say it.

The opening ceremony of the Emma Curtis Hopkins College was held at a sunny six in the evening in August 1995, in the school's in-progress library. By this time, the congregation of Unity-Clearwater had grown to almost 1,500 members. Eight had enrolled in the new college. The board of directors, their spouses and friends, and students and faculty gathered round as Leddy gave the welcoming invocation. "Dear friends, we bless and praise the grand opening of the Emma Curtis Hopkins College here this beautiful Indian summer," she said, beaming at the group. "This unique educational opportunity has a long foreground, and yet there is much more to be done by all of us who care so deeply about Emma Curtis Hopkins College, because of the possibilities offered future generations."

Leddy urged those present to give the project the encouragement of their spoken and written word, the welcome of their actions, the gifts of their monetary substance, and the power of their prayer. Each person received a keepsake Emma Curtis Hopkins College bookmark, the back of which bore a portion of Hopkins's own baccalaureate address from when she graduated the first class of New Thought ministers in 1891. "Although these words were first shared by the mother of the New Thought Movement, they echo in our hearts and minds today," Leddy said. She closed with a blessing, and offered refreshments to the group.

As the months added up, work on the library was diligent and accelerating. On Saturdays, volunteers listed and moved reference books onto two shelves in the cataloging room. Leddy and Russ, her husband and a former rocker in charge of the church's audio-visual needs, donated three large bookcases to replace the small metal ones—and the college board was requested to release funds for the purchase of five additional bookcases, which they obliged. A generous church member donated forty dollars toward the effort. When Dell's mother passed, he bequeathed her collection of hundreds of theological texts to the library: Unity books, a New Thought collection, and Christian Science books and journals. Space was made for more shelves to be built—all believed funds for a cabinetmaker would be forthcoming very soon. The board unanimously voted to call the library deChant Library, in honor of Dell's mother.

U-P.C. satellite centers began to open up around the county. Charles Throckmorton and his partner, Brent, opened a storefront church on Madeira Beach, which my parents and I soon began attending with Dell's and Leddy's blessings, in order to support the spread of truth. My parents continued their work with the U-P.C. and the Emma Curtis Hopkins College. The boards of both missions moved forward with great faith, knowing Divine order was always at work in their endeavors.

III.

When I was twenty-six, I eloped to California with a man I'd been casually dating for less than six months. He was an acquaintance from college—we'd traveled in the same group but rarely hung out. After graduating, he moved to California and I moved to Florida, and after reuniting with him on Facebook, I discovered he was also a writer. I asked if I could publish one of his short stories in a small journal I was editing at the time. We became pen pals. Every few weeks, a piece of local trash from our respective coasts would arrive in a mailbox: takeout menus, receipts for coffee, matchbooks, lost homework. He typed up Biggie lyrics on a typewriter. I sent him a shell, black and big as my hand, in a cigar box. That August, we moved back to New York at the same time and started hanging out every day. On New Year's Eve, we kissed for the first time.

As a child, my husband was a Mass-frequenting Catholic. He went to Catholic school for seventh and eighth grade and then Jesuit high school, and in Jesuit school, they went to Mass. The discipline of it instilled in him a certain pride, but he also took from it a conviction of God's vindictive nature. He believed he needed to earn God's approval to counter his own base impulses, and that if he followed his gut instincts, he would likely lose it. He still has a difficult time trusting in his successes or believing he deserves them. He apologizes much more than I do. He's believed, at times, that forces beyond his control were conspiring against him. And that people didn't like him because of some intrinsic trait.

He also believes that voicing a fear makes it less frightening. In Sunday school, he went to confession weekly. On the way, if he couldn't think of how he'd sinned that week, he would instead come up with things to lie about and say he did wrong. The consequences seemed arbitrary, that each sin could be resolved with a requisite number of prayers. The process was nerve-wracking and nonsensical. But after my husband spoke his sins, real or imagined, aloud, the priest would absolve him of them, and my husband would be freed. He could walk away confused but feeling once again safe.

When something goes wrong, my husband believes in the worst possible outcome. If we're low on cash, we're going to starve. If someone doesn't answer an email, they're avoiding him. Doing something wrong the first time he tries it has sent him into a panic. His tendency is to speak his fears aloud because in saying them out loud, he can hear how ridiculous they sound. Then they're not scary anymore—by sharing them with someone, he's able to let them go, even if fleetingly. Preparing for the worst, to him, is preferable to hoping for the best and being disappointed—in which case he's the fool. When we argue, he tends to concede I'm right and that his list of sins is endless.

I once heard him say, while sending a film festival application, "I won't get in, anyway. They always choose the same people."

"Don't say that," I said. "How can you not get in? You're the best."

"You know I'm not."

"Stop saying bad things about my husband," I replied.

"I'm not," he said. "It's the truth. I'm not sad."

He smiled to show me he was fine.

When my husband and I were a few years into our marriage, I found myself feeling irrationally irked at him over minor incidents. We typically get along easily—we're best friends. We make each other laugh all the time, sing together, feed each other. We can't go three hours without talking. He's the funniest, sweetest person I know. But we had gotten into a cycle of repeating the same argument over and over again: he would say something negative, and I'd jump down his throat. It got to the point where even relatively neutral statements about the laundry taking a long time or the refrigerator not being cold enough sounded negative to me.

I finally realized his tendency to articulate negative ideas and fears was disturbing me. For me it was beyond his pessimistic tendencies and my optimistic ones—in articulating his pessimism he was sharing it with me, and in listening I was committing the mortal sin of casting doubt.

By that point, I hadn't been to church in seventeen years—I had stopped going entirely around the same time as my parents, when I was twelve, and had forgotten a lot of it. My most recent memory was of the storefront satellite ministry founded by Charles and Brent on Madeira Beach. We had never returned to Unity-Clearwater after that, and before long, the satellite ministry closed. Unity had become a distant memory, one inseparable from my childhood.

I remember standing in the kitchen of my childhood home,

practicing affirmations with my father. Shouting, "I'm not stupid!" in frustration—him responding, with his infinite patience, "I am smart." I remember watching my words, afraid that I might affirm something bad accidentally, that I would fail a test or hurt someone I cared about. Over time, it became a kind of superstition in which I believed admitting I was afraid, even acknowledging the feeling to myself, would cause my fears to manifest as reality.

Doubt to me was equivalent to mortal sin: to cast doubt upon a hope was to do away with all possibility for that hope's fulfillment, and would doom me to a life of hopelessness. I refused to live a life of hopelessness. I was a child of God—and therefore I was successful, I was prosperous. I repeated this silently to myself. I believed in the power of I Am.

Despite all the time that had passed, despite the fact that I had long forgotten Unity's place in my life, in my mind, when I married, positivity wasn't doctrine; it was truth. It never occurred to me that the yin might have a yang, that someone could find safety in negativity rather than positivity—negative outcomes were not possibilities I even wanted to acknowledge. Hearing them voiced so often by my husband infused my life with a low-grade terror that manifested as irritation. In saying bad things out loud, he was making them real—so I had to make him stop. The more I tried to explain this to him, though, the less it made sense, even to me. The cause wasn't matching the effect.

I return to what is now the Unity Church of Clearwater with my mom in the summer of 2015. The sanctuary's tan-backed theater seats have been replaced with green upholstered ones, and the brown carpet with beige. Where rust-colored curtains once hung upon the stage alongside the gold-winged globe, there is an impressionistic painting of a seascape and wide white walls flanked by screens displaying soothing nature stills. A man sits at a drum kit with his sticks in his lap; another with a ukulele stands near him, and a woman with a guitar, with placid, faraway looks in their eyes. Above our heads, people murmur in a media loft, conferring over technical issues—I recognize Leddy's husband among them. The room fills and people greet one another as they take their seats. My mom and I sit a few rows back, on the aisle.

At ten thirty, three high-pitched tones play over speakers built into the ceiling and the doors of the sanctuary close. The room falls quiet. A middle-aged woman approaches the microphone and sips from a bottle of water, smiling at the audience. "Prayer is the heart of our Unity ministry," she begins. "Prayer request forms are found in the lobby and in the prayer room down the central hallway."

On the screens, a message appears in white above a babbling brook:

MANY FIND GOD IN THE SILENCE. LET'S BE STILL AND HELP THEM IN THEIR QUEST. THANKS FOR TURNING OFF YOUR CELL PHONE.

"Let us affirm Divine guidance, healing, prosperity, freedom, and peace for every name that has been shared as we begin our worship with a reading of today's Daily Word lesson," says the woman. "Take a deep breath as we become centered."

The message changes on the screens: "Rejuvenate." And changes again: "I am whole, strong, and full of vitality. Nature in its summer splendor invites me to rest and rejuvenate. Cats and dogs bask in the sunlight. A gentle breeze carries the scent of flowers. The sunset provides a moment of splendor at the end of each day." And changes again: "You have been born anew.—1 Peter 1:23"

"It's peaceful, I guess," my mom jokes.

The woman leads us through a guided meditation on the Daily Word and then steps to the side as the man with the ukulele takes over the microphone. His hair is middle-parted and very clean, almost fluffy. We sing along with his original song, written around the theme of rejuvenation, following the lyrics as they appear on the screens. Afterward, he leads us in a rendition of Unity Church of Clearwater's anthem, "House Built on Love."

"I don't remember it being this corny," says my mom.

Leddy comes up the aisle as we reach the final chorus. She shakes hands with people here and there, smiling solicitously, her expression a perfect example of tranquility, her red hair down and cascading over her shoulders. She reaches the microphone just as the song ends. A live feed appears on the screens.

"You are here by Divine appointment," she says.

"Life bringers" deliver Unity Church of Clearwater tote bags to newcomers, who make themselves known in the audience. Leddy begins her lesson. The topic is family. "How do we get to be a family?" she asks. "We came across the universe to find this cluster of souls."

She speaks on this topic for a moment and then calls the ukulele man back to the microphone. He tells the story of his own family: he was adopted when he was eight from the Big Brothers Big Sisters program along with his two brothers. As he launches into another song, this one telling their story, a woman in the second row lifts her hands into the air and leaves them there for several minutes.

I lean over to my mom. "She can't put her hands down now," I whisper. "She just has to leave them there."

"She's starting to wobble a little bit," my mom whispers back.

The song ends and the woman lowers her hands. Leddy returns to the front. "There are no accidents," she says. "Every child chooses to be born."

She steps down from the stage.

She tells the story of her own first child, who was born when she was a teenager. After the delivery, a social worker visited Leddy in her hospital room. She lied to Leddy about the paperwork she was asking her to sign. Leddy's son was taken from her. He found her again after twenty-five years—he'd been adopted by Mennonites and knew nothing about her. Leddy asked him whether he hated her. "You gave me life," he told her. "Of course I love you."

By now, Leddy is crying.

A message appears on the screens: "I am a child of God and therefore I do not inherit sickness."

"You claim your legacy," Leddy says. "As much as you would like to spare your children any sorrow, they have to claim it for themselves. I do believe in the reincarnation of family problems—but your mission in life is to learn to love them."

The audience stirs. "Leddy is keeping it real today," says my mom.

Leddy tells the story of her half brother, who collected money from the neighbors when his younger brother was in a motorcycle accident, then spent the money on drugs.

"Hurt people hurt people," she concludes, "and free people free people. Freedom is a choice."

The ukulele man returns to the front for another song as the lifebringers circulate wicker baskets. We say the Prayer for Abundance and close the service in the Unity-Clearwater tradition with "Let There Be Peace on Earth," all of us lifting our hands into the air for the final line: "Let it begin with me." A karaoke version of "House Built on Love" plays over the speakers as we exit the sanctuary.

Leddy finds my mom and me in the common area. We sit with her on an upholstered bench beneath a gold-framed mirror, sharing news of our families and life changes over the past seventeen years. I ask her when they remodeled the church—I'm excited to see it again, I admit, but it seems a lot has changed. She tells us about the fire that had gutted the original Unity-Clearwater in May 2003, four years after my parents and I left the church. A cable had sparked in the seminary library. The bookshelves ignited and the fire ate through the church's central corridor, into the prayer chapel, the fellowship hall, the sanctuary, and the youth ministry, through the offices and the meditation garden, narrowly missing a nest of duck eggs that Leddy saved and later hatched in her garage. What wasn't damaged by fire was damaged by smoke. What wasn't damaged by fire or smoke was laden with water. Everything except the shell of the building was destroyed.

They held Sunday services under a tent in the parking lot while the church was being rebuilt. There were two each Sunday: one in the morning and one in the afternoon. Both times, the tent was filled. But the church lost two-thirds of its congregation. "We survived in spite of the fire," Leddy says. "And now I want to say, in cooperation with it."

It enabled the church to make the leap into new media. They built the loft in the sanctuary and started live-streaming Sunday services on the website. Now online donations fund the majority of the ministry.

"In New Thought, there are no accidents," she says. "We pride ourselves on seeing the relationship between cause and effect. That's the qualifying essential of intelligence."

I ask what faith's role is in that case.

"I'll give you the Unity catechism," she says. "'Faith is the perceiving power of the mind linked with the power to shape substance.' Those are the words of Charles Fillmore."

Leddy once explained to me the mechanics of the universe. She used the book of Genesis as an example: "'In the beginning . . . God said let there be light, and there was light,'" she told me. "In Hebrew, the better translation would be 'In beginning anything': 'In beginning anything, God said, Come, light.'" He called to the light itself—called to the light's potential. Not planets, not stars, but their very essence. We first call for the potential of the world, its idea, and then we create it. I still believe this.

My mom's favorite New Thought writer is H. Emilie Cady. Cady was a homeopathic physician in New York when Myrtle Fillmore discovered her pamphlet "Finding the Christ in Ourselves" and asked her to write a series of articles for *Unity*. Myrtle and Charles later collected Cady's writings as *Lessons in Truth*. It was the first book they published on behalf of the Unity movement, and it quickly became the movement's most important text, second only to the Bible. Organized into twelve chapters—now a hallmark of New Thought books—it lays out the foundations of New Thought philosophy and its methods for healing. At the time of its publication, it was considered radical—unprecedented within the movement. It has since sold millions of copies and been translated into twelve languages. Maya Angelou once called it a revelation on *The Oprah Winfrey Show*, crying as she told Oprah, "God loves me . . . It still humbles me."

Cady's writing is clear and concise, relatable and inspiring, even moving. Reading it now, I find myself underlining and taking marginal notes on almost every page.

God is principle, she says, as music is principle. To find God, we must simply look within ourselves. We cannot find our Father-Mother in a book—even hers—for books can only give the reader another's opinion about Truth: "Seek light from the Spirit of Truth within you. Go alone. Think alone," Cady says.

God is Good and therefore all is Good—that is God's principle, and therefore nothing can be evil. "The sun does not radiate light and warmth today and darkness and chill tomorrow; it cannot, from the nature of its being," she says. "We do not have to beseech God any more than we have to beseech the sun to shine."

I asked my mom recently whom she believed she was praying to when she still prayed. She said it wasn't so much that she was praying to a being, but to a life force running through everything. It was really more about how the universe functions, she said. She repeated the same truths that grounded her work as a victim advocate all those years ago: she still doesn't believe in pure evil; she believes that people can behave in an evil fashion, but that they came by that behavior through their own life experiences.

Cady says evil is but an error perception of our mortal mind—as separate from our Spiritual mind and as separate from our body, which is itself an expression of God, the least part of Him. To remove the appearance of evil from your life, you must cleanse the mortal mind of error thinking by denying its existence. "If you repeatedly deny a false or unhappy condition, it loses its power to make you unhappy," she says. This is where Cady's thinking falls apart for me.

She says, "It is perfectly natural for the human mind to seek to escape from its troubles by running away from present environments or by planning some change on the material plane. Such methods of escape are absolutely vain and foolish." But what if your partner is beating you? Is escape then still vain and foolish? Of course not.

She says that "victory must be won in the silence of your own being first," and I think of my mother's experience of finding the indwelling Christ while Bob was hitting her. She says, "You need take no part in the outer demonstration of relief from conditions."

I picture of my mom putting back the knife, hiding in the school yard, curled up in a ball as the blows fall.

Cady says that "inharmony cannot remain in any home where even one member daily practices [an] hour of the presence of God, so surely does the renewed infilling of the heart by peace and harmony result in the continued outgoing of peace and harmony into the entire surroundings," and I think of a woman named Linda Osmundson.

My mother worked quite a bit with Linda Osmundson when she was still director of the Spouse Abuse Shelter, and doing advocacy work with the Domestic Violence Task Force and the Florida Coalition Against Domestic Violence. Linda was director of a shelter in St. Petersburg called Community Action Stops Abuse, or CASA, until she died in 2016. She was a fierce woman—one of the bravest I've ever seen. She went up against powerful opposition to protect victims of domestic violence. She was also a survivor herself and a lifelong Christian Scientist. I asked if I could interview her about her faith and work, and she invited me to her home.

It was difficult to understand Linda during our interview, and her sister often had to translate for her. A large bandage covered the left side of her face, on which grew a tumor that, I would estimate, weighed five pounds. It significantly impeded her speech. The year before, her dentist had discovered a growth inside her mouth. She sought treatment from a Christian Science practitioner, as she did not believe in conventional medicine. She did not publicly recognize the tumor—neither of us acknowledged it during the interview.

Linda was a domestic violence advocate for almost three decades. She had already been working at CASA for several years before she felt able to divorce her first husband. Most of the people she worked with didn't know she was married—her husband was in prison for eight years before she divorced him. And after that, it took her many years to tell people about her experience of abuse. She didn't want to damage CASA's credibility.

"Unfortunately, my faith kept me in the relationship too long," she told me. "I believed I could heal the marriage."

I asked her what, in her opinion as a Christian Scientist, the role of language was in domestic violence recovery.

"I don't use the word 'recovery,'" she said. "It's an addiction word that we have to really disassociate ourselves from. I don't see battered women as sick or crazy. I see them as, something happened to them. It's not something you did, it's not something you caused."

True: my mom's wrong thinking didn't cause Bob to hit her, and Linda's wrong thinking didn't cause her husband to hit her, and Linda's wrong thinking didn't give her cancer. But it occurred to me that recovery encompassed more than Linda believed. We don't only recover from mental sickness, but also from physical illness and injury. Even if we don't cause our own pain, we still have to recover from it.

I asked Linda whether she believed a world without violence is possible. "I don't know if it's possible," she said. "I go forward with the idea that it's possible."

"That's Christian Science," said her sister.

"If you don't have hope, you can't live," she said. "Without hope, there really isn't any reason to be. That's where the faith comes in. Even if I don't understand it right now, I've seen enough proof of it that I know it will be possible."

My mom and I went to Linda's memorial service together in January 2016. It was held in the historic Palladium Theater in downtown St. Petersburg. Hundreds of people attended. People from the community and the shelter gave eulogies, as did her sisters and her second husband, Maurice, a Quaker and fellow bicycle enthusiast Linda met in her riding group just a few years before.

Near the end of the service, an advocate from CASA read a quote that has been hanging in my mother's office for as long as I can remember. The version on her wall was printed on a background of Matisse's *Blue Nude*, a paper cutout of a woman curled up in a ball with her right arm covering her head. The quote is attributed to an anonymous domestic violence survivor. It says, "So I fight with one hand and love with the other. In some of my dreams though, I love with both hands, and the fighting is over."

My mom's papers from the Emma Curtis Hopkins College and Theological Seminary end in 1999. That year, my dad resigned from the Unity-Progressive Council board over interpersonal conflicts, and we joined Charles and Brent in their storefront satellite church on Madeira Beach as a way of supporting their spread of Truth. Twenty other Unity-Clearwater congregants came with us. Now, instead of theater seats, we sat on folding chairs. Our hymnals were laminated photocopies of Unity songs. It was hard to grow the congregation. After a few months, Charles was forced to shut the doors, unable to pay the rent. We never went back to Unity-Clearwater: the culture had changed—my parents felt there was too much negative thinking in the church, and too many power struggles. As my mom puts it, "We were church homeless."

In the years since leaving Unity-Clearwater, my dad has been forced to confront the logical inconsistencies of his faith: why horrible things happen to good, faithful people; why evil, conniving people advance in life; why suffering so often accompanies death; why his own father suffered for years as he died when he was a good man who did good things for other people his whole life. If people thought rightly, and believed rightly, and held positive opinions, and were good to one another, my dad said, then, according to Unity's tenets, the outcome should be good results.

"I saw no evidence of that," he told me. "If the principle is a principle, the principle should perform. I began to say to myself, 'You know, Eric, if you are a thinking person, you have to accept that there is no guiding spirit, no guiding force, no guiding principle in the universe insofar as it would lead to God, or to a god.' The belief is nothing more than a subtler form of belief in magic." Today, my dad is an atheist.

When I asked my mom how she defines her beliefs, she said, "I don't know if what I believe in could be called any kind of deity. I believe there's a spiritual element to all of us, whatever that life force is."

We were talking on the phone, something we do more often now. We've grown closer as I've grown older. She's become more than a myth. There was a pause as she chose her words carefully.

"When someone is dead, you can tell it's gone," she said. "I think somehow we're all connected, by being spirit. Our cores. We're all struggling to find some meaning."

Going Diamond

Sections of this essay are fictionalized composite accounts based on tours in the Bayou Club and Feather Sound communities between July and August 2015.

I was seven when my parents joined Amway. Our house filled up with Amway products: boxes of Nutrilite vitamins, Toaster Pastries, Glister toothpaste, Artistry makeup. We washed our hair with Satinique shampoo; we washed our floors with L.O.C. cleaner; we washed our dishes with Amway-brand dish soap; we strained our drinking water through Amway's filter. Our friends were Amway. Our vocabulary was Amway. We were "Directs" going "Diamond." We showed "The Plan" to anyone who listened.

We drove to Miami for "functions" at the Fontainebleau Hotel. Thousands of people attended, all packed into the big ballroom with lights turned up and people dancing in the aisles, getting "fired up" to Calloway's "I Wanna Be Rich," which blasted over the speakers. We clapped our hands and sang along.

A man took the stage with a microphone—a *Diamond*!—followed by a woman in a ball gown—another Diamond! Another Diamond and another and another, all shining under spotlights, smiling—their success itself a luminous aura engulfing them. "DO YOU WANT YOUR DREAM TO BECOME A REALITY?" the man yelled, strutting and flashing his teeth. "WHO'S GOT A DREAM?"

We had a dream!

"I SAID, WHO'S GOT A DREAM?"

We did!

We drove our teal '88 Oldsmobile Delta to the Bayou Club Estates for our requisite "dreambuilding" and toured the brand-new houses: big mansions with tall, echoing ceilings and screened-in pools, shiny state-of-the-art kitchens, garages big enough for three Mercedes, a golf course in the back, vanity mirrors and crystal fixtures in every bathroom. We drove to the yacht dealer and toured the Princesses and the Prestiges, lying on cabin beds and ascending the wooden stairs to stand on pulpits, gazing toward imagined horizons.

Amway is a multilevel marketing corporation. Some call it a pyramid scheme. In 2015, its parent company, Alticor, claimed transglobal sales of $9.5 billion.* It is the biggest direct-selling company in the world. Distributors make money by signing up other distributors and—somewhere in the background—"selling" Amway products. It's not exactly clear how Amway products should reach the public. That isn't part of Amway's marketing plan; The Plan mostly teaches distributors how to sign up other distributors, to whom they then distribute Amway products, who then distribute Amway products to other distributors they sign up, and onward. Amway has been the target, along with its affiliate

---

* This represented a second year of decline for Alticor, whose sales in 2014 totaled $10.8 billion.

companies, of multimillion-dollar lawsuits and other legal actions on almost every continent.

Four years after joining Amway, my parents came to their senses. There was L.O.C. cleaner in our closet for years while we pretended Amway never happened.

But every time I drive past the Bayou Club, I can't help wondering what it would have been like to go Diamond. Once considered the highest Pin Level—above Silver, Gold, Platinum, Ruby, Pearl, Sapphire, and Emerald—Diamond status was what I had craved. It was what I'd believed was success. After all, less than 1 percent of Amway distributors go Diamond.

SILVERTHORN ROAD, SEMINOLE, FL 33777

4 BED, 4 BATH, 5,144 SQ. FT.

$725,000

We've gone Diamond. "We're buying a house in the Bayou Club. We're starting a family," we tell the Realtor.

The first we see is in the Estates section. Croton in the front yard, Alexander palms and twisting cypress—all yards are maintained by the Bayou Club's landscapers, she says. Each yard must coordinate with every other yard, to meet color-palette standards that coordinate with every house. You pay $137 a month for this privilege, another $205 for security and maintenance of common areas.

This house has two stories, an office and a loft, bamboo floors, a three-car garage, a pool.

"You can see we're getting the screens fixed," the Realtor says, pointing to the men working beyond the glass. She has piercing blue eyes. Processed blonde hair. She has French-tipped nails, diamond rings on all fingers, and a gold-and-diamond necklace. She wears a white semisheer shirt and black-and-white-printed leisure pants, black eyeliner and heavy mascara. "We're just putting some finishing touches on the place."

I approach the French doors. The pool is bordered by stocky palms and, beyond them, the twelfth fairway. There is nothing like a yard.

"Can children play on the golf course?" I ask.

"No, it's private," she says. "And unless you don't love your children, you don't let them play on the golf course because they'll be golf ball magnets."

My husband chuckles.

"And the golf course is private," she says again. "You have to join the club. If a golfer sees a child out there running around, they will call the golf ranger to chase them because they interfere with the game of play.

"Do you play?" she asks my husband.

"No, but family does."

"We pay for golf privileges and we don't like people on the golf course. We like our fairways nice and even."

I wonder where the children play. The front yard is tiny. There's barely any grass.

"So the kids play out front," she continues. "And you know what? They do. When a child goes outside, he brings other kids out. We're very strict about our speed limit here."

"I noticed the speed bumps," I say.

"There are no speed bumps," she says, and I feel embarrassed. "If you came through Bardmoor, next door, there are bumps, but there are no bumps in Bayou Club. A lot of people have low-profile cars. We control our speed through our rover, who shoots radar. The fines are strict."

"Is there a neighborhood watch?" asks my husband.

"We have two security guards: one that roves the community 24/7 and one that stays at the gate," she says. "It's not a hundred percent safe because if somebody wanted to come through Bardmoor, hop that fence in the middle of the night, and intrude on your house, nothing's going to stop them. That gate out front is not going to stop them."

"It's hardly even a gate," I say.

"Your car won't get through it," she says. "They might steal your jewelry, but they're not stealing any big items."

"It has the illusion of security," says my husband.

If it's not your family who brings you in, it's probably a friend. For my dad, it was a manager at one of the car dealerships for which he handled advertising. The man's business comprised almost half of my dad's income. Over time, they'd developed a friendship. You'd think my dad would be immune to Amway, given his familiarity with advertising's insidious ways. But how does the saying go? A good salesman can sell you your own grandmother.

My parents and I were solidly middle-class when we collided with Amway. We owned our home. We lived in a safe neighborhood where I could play outside without supervision and walk home alone after the sun went down. We always kept an excess of food in the house. I got new shoes whenever I outgrew my old pair. I received new toys when my old ones broke and new books when I finished reading the ones I had. I went to gymnastics practice four times a week, singing lessons once a week, camp over the summer, and back-to-school shopping in the fall. We didn't need Amway.

But that didn't matter. In Amway, there's no such thing as contentment.

If you're happy with what you have, you haven't dreamed, says

Amway. Your life could be faster, shinier, brighter, more spacious—don't settle for less. Join Amway.

You could drive a Jaguar instead of your crappy Oldsmobile. You could build a custom home—don't settle for that two-bit shotgun you have. If you're proud of what you've accomplished so far in your life, don't be. Think bigger. Do better. If you don't believe you can—trust Amway. Amway believes in you.

Nothing was wrong with our life before Amway—we didn't join it to fill a void. We were happy, until we were told we could be happier.

Amway: The True Story of the Company That Transformed the Lives of Millions reads like an extended advertisement. Its author, Wilbur Cross, became acquainted with Amway cofounders Rich DeVos and Jay Van Andel when they commissioned him to write the first "official" history of the Amway Corporation, *Commitment to Excellence*, published in 1986. In *Amway*, Cross repeatedly references the work of Shad Helmstetter, PhD,[†] a "motivational expert" specializing in "programming" yourself to change negative self-talk into positive self-talk. Negativity is expressly verboten in the world of Amway, as it breeds doubt—distributors are advised to get rid of any negative people in their downline as soon as possible if they can't train them to be positive.[‡]

[†] He studied at Harper College in Palatine, Illinois; San José State; and Southwest University, a correspondence school, where he received his PhD.

[‡] There are long sections in *Amway* dedicated to teaching readers how to go about programming others to be positive—you are helping people in this way, Cross says, "You can offer them hope when they feel as though they are in a hopeless situation. Best of all, there are no limits to those you can help."

Helmstetter credits the practice of "dreambuilding" as a central reason why Amway is so successful. Dreambuilding is more than wishful thinking, Cross explains. It's more than seeing what people with more money have and wishing you had it. Dreambuilding is "the perfection of excellence"—"It is a way to control what you *think*, to enhance what you *believe*, and to solidify your *attitude*" (emphasis his own). Most importantly, it's a procedure, "a skill that has to be learned, practiced, and put into action."

Put into action—that's where Amway comes in. It gives its distributors dreambuilding guidelines and opportunities, chances to "practice the art . . . on a daily basis."

We slipped Amway motivational tapes into our car's tape deck, and listened, and repeated. We bought tickets to Amway functions for fifty dollars a pop and booked hotel rooms nearby to attend them. We sampled the products and demonstrated our commitment by filling our house with boxes upon boxes upon boxes of Amway goods.

We made lists. We framed pictures. We drew diagrams. We hosted seminars in our home where we lectured our downline to "activate their dreams." We constantly reminded ourselves that our dreams were possible. We only interacted with others who affirmed this.

In fact, Cross says, it may be in your best interest to seek out the most hopeless and lonely people you know and invite them to become Amway distributors. Sometimes "negative people turn out to be the best prospects of all" because they are the ones who really need Amway most—"they are the ones who are looking for something else in life." Likewise, you should "locate the lonely individuals of this world and offer them something that will counteract their isolation."

We took photographs of one another inside our dreams: Here I am, a skinny nine-year-old posing proudly next to a kidney-shaped pool. Here's my mother in a pair of khaki shorts and a Hawaiian shirt descending a marble staircase. And my father, two thumbs up, lying on a king-sized canopy bed. We visualized, yes—but then we went one step further and *made visual.* We stepped inside our dreams, literally.

Dreambuilding is Amway's profit engine. Tour the house for motivation—but then how do you buy the house now that you want it? Buy tickets to the Amway-hosted functions. Buy training tapes and manuals sold by upper-level Amway distributors. Build The Business. Prospect others. Buy Glister and Satinique.

Building dreams was like building a house: it wouldn't work with wishes alone. Wishes were ephemeral; we needed something concrete. In order for our dreams to feel real, we had to construct them from tangible things. The Plan and The Business were our brick and our mortar.

WE STEP INTO THE SPACIOUS KITCHEN. IT HAS A WRAP-AROUND GRANITE COUNTERTOP, STAIN-less steel appliances, beige tile and smoke-colored grout. I do a spin to peer into the breakfast nook, decorated with gauzy floral curtains and a chandelier.

"Most people think the kitchen needs updating," says the Realtor.

"I like it," says my husband, placing his hand on the small of my back.

"You'll want to rip out the kitchen, replace the flooring, put your own touches on the home," she says, giving me a wink. "Make it your home."

She takes a folder from the counter and opens it before us. Inside is a floor plan of the house, a map of the community, an elevation certificate, and a packet listing features of the house, with accompanying photographs.

"Not every Realtor does this for their listings." She points to a spot on the map. "Bayou Club is divided into three communities. Here is where this house is located within the community—the Estates. And here is Sago Point, and here is Copperleaf."

"What's the difference?" asks my husband.

"There are four hundred single-family homes in Bayou Club," she says. "No condos, no townhomes—all single-family. Ninety of those homes are in Sago Point. They're not tract homes—they're different versions of the same home, and smaller: two thousand to three thousand square feet. Because of the size of the homes and the maintenance, they've attracted a lot of second homeowners and empty nesters. Somebody looking for something more children-friendly might move over to Copperleaf, where the homes are a little bit larger and the lots are a little bit larger. You may have three-car garages versus two-car garages. And then you can upgrade to the Estates section, where they're all custom-built."

"You could spend your entire life moving around the same neighborhood," I say.

"You could," she says. "Full circle. Bayou Club is very desirable. People want to live here. They may be living elsewhere and want to upgrade, and now they can afford to live in the Bayou Club after being in practice a couple of years. My husband is a physician, and I know you don't start out in the Bayou Club."

"What was this area before it was the Bayou Club?" my husband asks.

"It was very marshy. They rearranged the golf course because part of Bardmoor was in here, so they restructured it," she says, referring to the adjacent gated community. "Bayou Club is divided into two cities: Pinellas Park and Seminole. When you first drive into the community, while you're technically still in Pinellas Park, you wouldn't know it. Pinellas Park is low-income—we call this section an oasis in the middle of Pinellas Park."

We follow her up the stairs. There are two large bedrooms separated by a bathroom and a linen closet—the children's rooms. I step into the one on my left, which is smaller than I expected. It has wood floors and a closet with sliding mirror doors. Out the window, the neighboring house is less than ten feet away, and the space between is filled with broad-leafed palm trees. I hear the faint twang of the radio on the pool deck, playing "Sweet Home Alabama."

I turn around and step into the bathroom. I touch the faucet on the sink and lift the valve to open the water.

"Do you need something?" she says, turning away from my husband to address me.

"No," I say, shutting the water off. "Just testing it out."

IN HIS MEMOIR *SIMPLY RICH*, AMWAY COFOUNDER RICH DEVOS[§] TELLS THE STORY OF AMWAY'S origins. The country was in the last gasps of the Great Depression. Rich was fourteen. He was walking two miles through the snow to his high school each day, in his hometown of Grand Rapids, Michigan: wool collar popped high, galoshes squishing, wind in his face. Occasionally he would take the streetcar or city bus—but allowing time for the city bus meant having to rise long before the sun came up. "I needed more efficient transportation, and already being an enterprising type, I had an idea," he writes.

Enter Jay Van Andel, Amway's other cofounder. Jay had a 1929 Model A, which Rich had noticed both driving down his street and also parked outside his high school. "I thought a ride in this

[§] DeVos and his family are joint owners of the Orlando Magic.

car would surely beat the bus, a streetcar, or walking," says Rich. The rest is as saccharine as you would expect: good American boys working hard to make their dreams come true—an adventure full of family values and sturdy bootstraps with which one can pull himself up. It begins with the heartwarming story of their first joint business venture, running a pilot school, then segues into a comedy-of-errors trip on a sailboat—a typical masculine coming-of-age experience rooted in good old-fashioned American values like cooperation, perseverance, and leadership.

Rich and Jay go into business together selling Nutrilite vitamins, an early multilevel marketing scheme for which Jay's second cousin and his parents are already distributors. When Nutrilite goes kaput in 1948 after an FDA crackdown on their "excessive claims" regarding the products' nutritional values (about which Rich only says, "Until then, there had been no official government position on what type of claims could be made about dietary supplements"), he and Jay strike out on their own—the American way. They can do it! We know they can!

At the heart of Amway is the love of "free enterprise"—an equal-opportunity system in which determination alone is the path to achievement. If you have a dream, Amway says, and you try hard enough to achieve that dream and let nothing stand in your way, then success is guaranteed. That is the promise of what Rich DeVos calls "Compassionate Capitalism"—helping people help themselves.

Rich and Jay set up shop in Rich's basement selling Liquid Organic Cleaner, or L.O.C., Amway's first original product. With their trust in each other and the support of their loving wives, they're able to weather all bumps on their ride to the top, including the first federal investigation of Amway, by the Federal Trade Commission in 1975. In a chapter of his memoir titled "The

Critics Weigh In" (in Part Two, called "Selling America"), Rich says of the suit, "[We] considered the suit another government misunderstanding of business principles and an attack on free enterprise."

Anything that challenges Amway—particularly the government—challenges free enterprise, and thus freedom itself.

Of the Amway distributors who testified in the case, Rich says, "I have nothing against someone who tries Amway and concludes the business is not for them. But I wish they would take responsibility for their own actions instead of trying to blame the business." Likewise naysayers and disgruntled former Amway distributors simply do not understand how business works and are at fault for their own failures because they lack faith in their ability to succeed, and thus the necessary determination.

If you need proof of Amway's principles, just look at all the people who've benefited from Amway, Rich tells us: millions worldwide.

But don't take it from him; take it from distributors themselves.

In a YouTube video uploaded in 2011, two Amway distributors talk to each other before the start of an Amway function.¶ Bass blasts through the dim roar of the packed stadium, which is decked in American flag colors. Spotlights rove over the audience.

"We here, man," says a young black man in a blue T-shirt. "See all the IBOs. It's good to be with people in your company, to feel the love. A lot of people back home be wondering how it is and how big of an organization it is. You see: just imagine the potential of having all these people in one group, man, even if you get ten dollars off a person"—he points to a random person in the

---

¶ Tickets to Amway functions now cost between seventy and ninety dollars.

audience—"all these people. There's a whole lot of money floating around in here somewhere."

"This is not a scam," says the person behind the camera, pointing it now at the empty stage. "Everybody think it's a scam. Come on, man. If these many people got scammed out? I don't think so."

"Besides," says Rich of the birth of Amway, "we knew what we were really selling was an opportunity for people to succeed on their own and help others do the same through a unique marketing system. All it took was the willingness to work hard to achieve a dream."

That's all it takes? Sign me up.

EAGLE POINTE DRIVE, CLEARWATER, FL 33762

5 BED, 4.5 BATH, 4,466 SQ. FT.

$1,150,000

We've gone Diamond, we tell the Realtor. We're buying a house in Feather Sound. We're starting a family.

This one is just beyond the gate when we enter the neighborhood. It's desert-colored with a terra-cotta paving stone roundabout drive and another gate that retracts when we enter the code. There are two palms planted on either side of the porch, two more on either side of the yard, and another in the grassy area encircled by the roundabout. A row of perfectly rectangular hedges lines the front of the house beneath the picture windows.

There are five bedrooms, five bathrooms, a game room, a study, a fitness center, a spiral staircase and elevator, a three-car garage, vaulted ceilings, and a three-tier waterfall pool.

The Realtor is standing beneath the porch with a younger woman, waiting to greet us. They're both dressed in red. They smile broadly as they step onto the drive.

"This is Renata, my assistant," says the Realtor, motioning for Renata to extend her hand. My husband takes it first.

"Nice car," says Renata. We've driven here in a Porsche.

The Realtor compliments my husband's name: she knows two others who share it. "It's not a common name," she says.

"No, it's not," he agrees. "It's a family name."

Tuscan-style columns flank the entryway. Inside, the walls are painted with tacky murals of Venetian street scenes. The elevator at the edge of the living area is cylindrical and see-through. An enormous aquarium occupies the wall adjacent to the fireplace.

"You'll see as you go through, they went to Italy on vacation," the Realtor says. "When they came back, they tried to incorporate as many things from Italy as they possibly could."

"I love Italy," says Renata.

We follow them toward the master bedroom, passing the dining room.

"That is a very expensive chandelier," says the Realtor, pointing to the one above the dining room table. "I believe it's Murano. Twenty-five thousand dollars." My husband studies it.

A sliding glass door leads from the master bedroom out to the pool and a lakefront view with a dock. I open the door nearest me inside the room.

"Now, don't freak out—that's just your shoe closet," says the Realtor, touching my arm.

"I was going to ask you where I should store my wardrobe," I say.

She smiles maternally. "That one's just for shoes."

Renata tells us about the best local attractions, recommending particular farm-to-table restaurants and yoga studios as my husband and I make slow, opposing circles around the room. We meet in front of the master bathroom. The shower is wide enough for three people with three showerheads, a knee-high tawny-colored tile wall, and the rest of the walls completed with glass. The whirlpool bathtub could easily accommodate three.

"We could have fun in here," says my husband. I elbow him in the ribs.

We follow the Realtor back to the fitness center, a long room lined with rubber flooring, eight machines, and a large, flat-screen TV mounted to the wall.

"Renata, you know about the exercise room," says the Realtor, stepping back to let Renata inside. "I'm not an exercise person."

"Does the house come with the fitness center?" asks my husband.

"You can ask him," she says. "It's all depending on the negotiated price. He's willing to leave it all. He'll leave anything here that you want."

"Is that a tanning bed?" I ask.

"Yeah, it's a professional one, too," says Renata.

My husband lifts the lid. "Is there anyone in here? Maybe you could finally get a tan," he says to me.

"I'd fry in there."

"You could come in here, work out, and then go to the next room and do your work, and never leave the house!" says the Realtor, motioning toward the study next door.

"Sounds great," I say.

I loved the days when we'd go to the Bayou Club as a family. We began going immediately after joining Amway, when I was in second grade. The development was new, still under construction. There was space between the houses and the far stretch of the golf course undulating luxuriously around them. Model homes rose from the landscape like castles, bigger than any houses I'd ever seen—and vacant. Never occupied. Empty dreams, waiting to be filled.

As a child, I found the pleasure of being inside a big house to be endless. Future ownership had come to feel like a guarantee, so I took to imagining what life would be like in each one we visited. In this model of a girl's bedroom with its shelf of figurines, canopy bed with lace cover, pink painted chest, and carved mirror, contentment felt within reach. This room was assurance I'd never be lonely or bored; that I would always have something lovely to look at, and lovely things to say, and other children near me to validate my worth. I felt special, included.

Imagine watching movies in this home theater. Imagine riding an elevator up to my dream bedroom. Imagine swimming in this pool each afternoon. Eating meals in this dining room under an opulent chandelier.

Imagine having a whole closet just for pool toys. Imagine having a room just for getting dressed. Imagine having a refrigerator twice as big as ours. And an island counter. A bay window. A golf cart. A motorized walk-in closet.

Our house on Twelfth Avenue had three bedrooms; I wanted four. A house with four bedrooms came to seem normal; I wanted five. I wanted a library. I wanted a hot tub. I wanted a spiral staircase with a wrought iron banister, and a playroom, and a whirlpool bathtub, and a room just for practicing ballet, and a fitness center, and a poolside bar.

Sometimes we brought along a camera and took pictures of one another walking around the houses. We saw two or three in a day and then took the film to be developed. Back in our three-bedroom, we looked at the photos together, then stored them in fresh albums. In the photos, we wore the same outfits while the houses around us changed. We were the proud owners of three beautiful homes, the photos said—or this was one big home. One monstrous behemoth of a home comprised of three mansions smashed together.

Limitation on ownership was not a concept I was familiar with as a middle-class child—everything could be mine. I had never experienced a feeling of lack. I never wanted for anything I needed. I was never told we couldn't afford something I asked for. While the thing I asked for might be denied me, money was never given as the reason. "Spoiled" was a word I heard often from family and friends, and I was proud of it. I thought I deserved to be spoiled—I was fully ignorant of the negative connotations of the word. By the very fact of being me, I believed I deserved material things.

My mother grew up in a family that didn't have money: six siblings, and later three stepsiblings, with one working parent, the other a drunk. She loves to take me shopping. This was always reflected beneath our Christmas tree—but also throughout the year. She wanted her daughter to have things she never had.

Shortly after we left Amway, I remember asking my mother for bell-bottom jeans. It couldn't wait. I had to have them immediately. She was not in the mood to go shopping that day. Just wasn't in the mood. But finally, I convinced her.

I ran my mouth as we left the house, leaping down from the porch as I shut the front door. "I get all I want!" I sang.

I'm shocked by this.

My mother stopped in her tracks and ordered me back inside. The look on her face still brings me shame.

I wish I could say the story ends there, but it doesn't. A few minutes later, and after many apologies, we went to the mall.

The house is outfitted with an elaborate security system. A small room on the second floor holds the bank of monitors. There are cameras on every corner of the house, and at every outside door, and several around the pool. Three rapid beeps signal a door's opening. Even though Feather Sound is a very safe neighborhood, Renata says, and she never heard of any home invasions while she was growing up here, people are very particular.

"This would make a good nursery," says my husband, stepping into the room with the monitors.

I chuckle.

"Is it cooking?" asks the Realtor, glancing toward my stomach.

"Not yet," I say. "We're trying."

"Don't worry. It will happen."

"It's—"

"I know, it's hard not to worry," she says.

She has mentioned my parents' proximity to this home several times on our tour. We've told her we're moving back from New Orleans to be near them as we start our own family. It turns out her daughter now lives in New Orleans.

"She's a writer," the Realtor says. "She's a blogger and she works for me and a couple other people. She writes blogs, and manages our social media, and any other part-time job she can find. She says, 'Mom, we're all college-educated people, we just do different things! It's just so freeing!' And I'm like, 'What about a career?'"

"'Don't you want to buy a house?'" my husband adds.

"Yeah, she does want to buy a house," says the Realtor. "And I'm like, 'I'm not buying you that house!'"

We follow her down the staircase.

Outside on the pool deck, the rush of waterfalls drowns out conversation. We're forced to yell over it.

"Now, this is really a retention pond," she says, pointing to the body of water at the edge of the yard. "It's not a lake or anything."

"No watercraft?" says my husband. She shakes her head. "Not even a canoe?"

"Maybe a canoe," she says.

He walks past the sculpted concrete wall separating the yard from the water and to the end of the wooden dock with his hands in his Dockers pockets. The Realtor turns away from watching him to stare at me, likely searching my face for an expression of wifely pride. Finding none, she says, "There's a fire pit back here."

"Oh?" I say.

"And the Feather Sound Country Club is right across the street. You would have to join the club to be able to play golf. There are several clubs in the area."

"Yes, we've been looking into some," I say. "Like the Bayou Club."

"What's that?" says my husband, returning from the dock. "I like the seawall," he says to the Realtor. "Looks like it's steel-reinforced."

"Well, this is a unique little area," she says, ignoring his observation. "You've got your most expensive homes here. Most people like it because of its close proximity to Tampa and to St. Petersburg. If you're commuting to either place, it's pretty easy. Do you know where you're working?"

We tell her we both work from home.

"Are you a blogger?" she jokes as we follow her back inside. The noise of the pool is quieted by the shutting of the French doors. Renata is waiting for us in the foyer.

"We don't show a lot anymore for security reasons," the Realtor says over her shoulder. "The world is just not what it used to be."

"No, it's not," says Renata.

"I thought you might enjoy talking to Renata because she probably knows things you'd be interested in, in the area."

Renata opens the front door, to three beeps. We step back out to the front walk, leading down to the driveway. It's begun to rain lightly, polka-dotting the terra-cotta paving stones.

"I think International is the nicest mall in the area," Renata says, following the Realtor's suggestion. "You have Neiman Marcus, Nordstrom's, you have Free People—"

"J.Crew?" asks my husband.

"J.Crew," she says. "You have all the normal plus some specialties."

"I think I'd put a basketball hoop in right here," says my husband, regarding a patch of grass.

"How are you going to get into the garage?" Renata asks.

"Might have to add a few more garages," he jokes. "I've got a bunch of cars."

"You know, it's hard to find more than a three-car garage around here," says the Realtor, appearing behind us.

"We can keep some in the driveway," he says. "Get rid of some. We're only two people." He puts his arm around my waist.

"Just put the lifts in," says Renata. "That's what the people across from my mom's house do—it's really common, actually. Really common."

CROSS OPENS *AMWAY* WITH THE TESTIMONIES OF AMWAY DISTRIBUTORS FROM AROUND THE world. There is no country in the world where Rich DeVos's Compassionate Capitalism does not work, Cross says. "Listen to Alcimon and Marie-Chantale Colas."

"Listen to Sevgi Corapci."

"Listen to Nilufer and Merih Bolukbasi."

Listen to them. Listen to them. *Listen.*

"I was a salaried man working in a company for eight years," says Kaoru Nakajima, Japan's first Amway Crown Ambassador. "Now I am my own boss. Now I am free. Now I am selling products that make me proud. Now I am helping people in five different countries to build their own businesses. When I see so many people getting more abundant lives, I feel really excited."

Listen to Wu Dao-liang, an "old man" who joined Amway after retiring, rather than spending his days reading the newspaper and playing mah-jongg: "At the beginning when I started my Amway business, my family was worried that I could not handle it physically. But during these two years, they noticed I became more optimistic and more healthy."

Listen to Rosemarie and Otto Steiner-Lang, who joined Amway in the hope of funding their own construction company and now run their Amway business full-time: "We have found in Amway the independence we were looking for. This business is a doable and affordable solution for the problems in the labor market today. Amway, which represents free enterprise perfectly, postulates and promotes the initiative of the individual, reducing the burden on the public social system."

There are more than enough reasons to join Amway, it seems: the promise of being your own boss, the possibility of helping others, of making human connections, finding existential purpose, beating boredom and passivity, spending more time with family, earning the admiration of others.

Unfulfilled in your current job? Feeling trapped? Amway frees you.

Shy introvert? Learn to be more outgoing with Amway.

Bored with your life? Feeling adrift? Amway gives you purpose.

Live an overall happier, healthier lifestyle, with a low commitment and the potential to accrue massive wealth—for only the upfront cost of a training manual.

A housewife can do it. A retiree can do it. The physically enfeebled can do it. Those who have only known poverty can do it. Families can do it together.

In the roughly two hundred pages of Cross's book, however, there is virtually no discussion of how Amway actually works.

Among entire chapters dedicated to Amway's state-of-the-art manufacturing facilities and its pioneering move onto the World Wide Web, the "Amway Distributor Profile," its "Bootstraps Philosophy," and Amway's foreign expansion strategy, the closest Cross comes to summarizing Amway's business plan is in this passage:

> Attaining goals for greater success and profitability depends on each distributor's ability to sponsor other distributors, who comprise their "downline." Patience is a characteristic much required in this step because a distributor can advance in profitability and standing only to the extent that the downline distributors actually sell products and keep on generating volume.

How do they sell those products? Not in retail stores—Amway distributors can only sell their products directly to the public,** or to other Amway distributors. This may not seem so bad until one considers that the price point for many Amway products has been reported to be about twice that of similar products found in retail stores. Or that in blind tests, Amway products often score poorly.

So, how is it that Amway continues to profit? Cross quotes James W. Robinson, author of *Empire of Freedom: The Amway Story*:

> The export lifeblood of some countries is oil, for others it is cars, or diamonds, or food. . . . America's most precious export is not a commodity, natural resource, or manufactured product, but an *idea*: putting free enterprise in the hands of the common man and woman.

** However, door-to-door sales are discouraged.

Let us not underestimate the power of ideas. Cross provides examples of distributors who let nothing stand in their way. Just listen to the story of the Upchurch family, who persisted in Amway, making any sacrifices necessary, even after Hurricane Fran destroyed their home. Or the Janzes, who were desperately poor new parents with another child on the way when they learned that Amway was bigger than making money; it was a way to overhaul your lifestyle and live your dreams. Or Dexter Yager,[††] who didn't let a stroke stop him from achieving success with Amway and continued to operate his business at the same level even as he was learning to walk and speak again.

Listen to the story of Ed Johnson. Ed's family lost everything when the recession hit San Antonio in the 1980s. Ed's son showed him The Plan in 1992, and after some initial resistance from Ed's wife, soon the whole family was working hard to achieve their dreams the American way.

Then tragedy struck. Just as he was qualifying for Diamond, Ed had to undergo emergency surgery to remove a brain tumor. Then he had to undergo radiation therapy. Did Ed let this stop him? Of course he didn't. He "showed his mettle" and his "desire to get on with his life" by prospecting three doctors and six nurses while he was in the hospital recovering from brain cancer treatment—enabling the Johnsons to go Diamond sixty-two months after joining Amway.

"Although Ed's challenges would have devastated most fami-

[††] With his wife, Birdie, Yager was the first Amway "Founders Crown Ambassador 60 FAA"—or Crown Ambassador with sixty Founders Achievement Awards points—in the world, and is the topic of Stephen Butterfield's 1985 exposé book, *Amway: The Cult of Free Enterprise.*

lies, the Johnsons saw them as an opportunity to pull together," Cross says.

"There are no excuses," he quotes Ed as saying, "just performance."

BULLARD DRIVE, CLEARWATER, FL 33762

9 BED, 8 BATH, 9,498 SQ. FT.

$2,002,000

We've gone Diamond, we tell the Realtor. We're buying a house in Feather Sound. We're starting a family.

This one sits on a double executive lot. An artificial creek snakes around the yard. Flashes of yellow and orange spotted koi pass beneath our feet as we approach from the brick walkway. The house is split-level with two wings, a custom pool with cascading waterfall, billiard room, media room, workout room, steam room, six-car garage, state-of-the-art workshop, custom built-in bar, loft for quiet relaxation, hurricane shutters, large views of the golf course—and two bedrooms above the garage sequestered for the help.

"It's very dark," I observe. We've begun in the middle: a room with wood paneling, shellacked stone floors and walls, and a recessed circular area for entertaining, carpeted in emerald. Behind me, a pool table occupies most of a Turkish rug annexing the area beneath the open-style second-floor balcony. The Realtor stands near a grand piano and a stone planter housing ferns.

"Tastewise, it may not be your taste," she says. She is wary of us—I can hear it in her tone. "Or it may, depending. But this is a blank canvas, you know. If you really wanted to make your own touch, you can certainly do that."

"I'm not going to want to remodel a house while I'm pregnant," I say to my husband. "Or with a new baby."

"Let's take a look," he says. "No harm in that."

"It's too big," I say.

"Let's take a look."

He takes my arm. The Realtor smiles.

"The house is divided into two parts," she says. "This side over here is the original part of the house." She indicates the area behind us. "On this side, there's five bedrooms. And then there's this middle area, which is a big entertainment space. And then on the other side, you've got a very, very large master suite and another bedroom."

She leads us toward the original wing, descending three carpeted stairs, and proceeds down a narrow wood-paneled hallway, pausing at the entrance of each bedroom to invite us inside. The windows are covered with wooden shutters blocking out the light. A musty smell rises from the linens. We wander silently in and out of rooms.

"Seems like it's been a while since anyone did anything with the place," I say, pausing before a marble-patterned dresser built into the wall.

"Particleboard," says my husband from the bathroom.

"Well, back in the day . . ." says the Realtor.

"See, open this up," I say, unlatching the shutters. Light pours across the bedspread and a shelf of porcelain dolls. "Isn't that better? Why would you want to close it off?"

My husband appears in the doorway. "They don't like the light here," he says.

"You know, I had some friends who came out of New Orleans after Katrina," says the Realtor, leading us suddenly back toward the front. "It's funny, we go out and we have dinner with them and it still comes up."

We follow her into the kitchen. A set of wooden cabinets hangs above the island counter. Others wrap around the walls, enclosing the space. Each of the glass doors is etched in the corners with flowers. My husband flicks a light switch, only dimly illuminating the countertops.

"It has a sort of ski lodge vibe," he says. The breakfast nook is decorated with white-and-green floral wallpaper and a white crocheted tablecloth. "You could throw some parties here," he says with a smirk.

The Realtor passes between us and opens a glass door to three beeps. We step into a spacious screened-in porch holding several upholstered deck chairs and a chiminea, a level above the pool and overlooking the golf course.

"Through most of the windows and doors, you can see pretty much three fairways," she says. "Which is really pretty and serene. The water is not normally like this."

In the middle distance, a large body of water has overflown its banks and seeps outward into the grass.

"I thought that was a lake," I say.

"Usually you can walk right out to that golf course. I've never seen it like this. This is so full. The birds are loving it."

"Are there a lot of birds out here?"

"The pool deck is really very large and has multiple levels on it, so—"

"I didn't notice," says my husband. "Is this a shingle roof?"

"It is."

"The exterior really calls for shingles," he says. "You can't go and put a Spanish tile on this roof, the way the exterior is."

"And that is cedar siding," she says, looking from my husband to me. "People ask, 'Why is it discolored?' It's meant to do that. You can restain it if you want."

"I like the variation," I say.

"It doesn't attract termites," she says. "Termites will not eat cedar."

When I was ten, my parents bought a house for $200,000. My dad had been running his advertising agency out of the spare bedroom of our house on Twelfth Avenue, and when he hired his third employee, he set up a desk in my bedroom for the graphic artist to work at while I was at school. Then a neighbor called the city about all the cars parked on the street, and my parents cracked a plan to move into a bigger house and bring the agency into the new house with us. By that time, though, business had gone gangbusters, so it turned that out moving the company into the new house wasn't necessary, after all—my dad rented an office, instead. The new house was entirely ours.

It was a single-story, with four bedrooms, three and a half baths, a roundabout drive, and a screened-in pool. "You'll see the gates," I'd say to my friends when giving them directions to my new house, feeling endowed with importance, despite the fact that these were not real gates—they were only for show. "They're metal arches that say 'Carlton Estates,'" I'd say. These words tasted like gold. Carlton was a surname hyphenated invisibly after my own. I lived in Carlton Estates: that was surely worth something.

We had a fireplace, a poolside grill, and a river-rock deck with closing screens. We had an island counter. We had walls covered with mirrors. To get to my parents' master bathroom, I passed through a dressing area connected to a walk-in closet. The bedroom next to mine was expressly for guests; the one at the end of the hall became a study. One of two living rooms seemed intended only for show, and the planter inside the front door housed pots of plants—silk, they never wilted. The bathroom off the family room had an outside door and a shower for people coming in from the pool. We bought new furniture, new rugs, new artwork. I had never felt more proud.

The houses in Carlton Estates were a magnitude above those in our old neighborhood, where all of the concrete homes followed more or less the same design. These sat on larger lots and had deeper lawns, and each was entirely unique. There were second and third stories, and sloping, multilevel roofs. There were bamboo thickets obscuring homes from the street. Stone and wood exteriors. Stained glass windows. No sidewalks. No streetlights.

I would peer over the wooden fence wrapped around our backyard into the lives of those around us. A noisy macaw lived in a sculpted metal cage on the pool deck next door. At the house behind ours, a gazebo covered in vines sat next to a pool, larger than ours, which sat beneath an even taller screen.

Going door-to-door for a school fund-raiser, I walked the winding, Anglophile streets—Kent Drive, Kings Point Drive—that looped around to the Intracoastal Waterway and back again in a closed circuit. The farther I strayed from our street, the larger the houses became. One house looked like an old-time plantation. Another had a waterfall in the center of its circular driveway, and a bright blue roof. I stood in dark foyers and bright, airy kitchens, saw antique furniture and shiny out-of-the-box appliances and mysterious works of art.

At the top of our street, I found a squat house with no windows—they'd all been cemented over. The grass yard had been replaced with gravel. The pool was covered and the cover was filled with leaves.

At the bottom of our street, I found an iron gate with a code confining three mansions bordering the Intracoastal. I slipped through it and knocked on the door of the mansion farthest to the right. A man answered with an angry-looking English bulldog at his feet. "How'd you get in here?" he demanded.

The pool deck is made of a flexible material the color of sunbaked fiberglass and molded to look like stones. It bows beneath our feet when we walk over it, as if hollow underneath. I see my husband notice it, too.

"What is this material?" I ask, pointing to the deck.

"I've never asked and I probably should," says the Realtor. She's been telling us about the pool pumps.

"It's some sort of fiberglass or plastic composite," says my husband.

"It does flex to some degree," she says.

"I'd want to replace it," I say.

"It's better than those concrete pool decks they put in," my husband says. "I bet it looks beautiful from the golf course."

"It does," she says. "Have you seen it from Google Earth? It really gives you a perspective of where you are in relation to Tampa Bay."

My husband approaches the edge of the deck and looks out over the hedges and across the golf course. A flock of ibis has congregated around the standing water. I join him and see that he's smiling.

"You like this?" I say.

"I really like this," he says. He turns to the Realtor. "My father's a big golfer. I bet he'd sit right here and watch them try."

"There's only one little problem with this house," says the Realtor as we follow her back inside through a different door, "and that's 'Which door should I go in?' There are so many doors!"

We're back in the central area. From an adjacent room comes

the sound of a television and we make our way toward it. The room is ruby-carpeted with red-and-gray-striped wallpaper, three tapered wall lamps, and a giant projection screen angled downward. A man faces away from us in a floral upholstered recliner. He pauses the television when we come in.

"Don't stop for us," says my husband. "What are you watching?"

"*Law and Order*," says the man.

We commence watching television. The man is indifferent. At the commercial break, my husband asks the Realtor what kind of wood is used for the paneling on three of four walls in this room.

"Brazilian cherry," she says.

We move to the base of the stairs and she opens a closet to three beeps. Inside is a circuit board and a mass of wires. "This house obviously has a lot of lines and cables in it," she says. "This is the command center, the central nervous system." She shuts the door again.

At the landing of the stairs, she turns to face us. "The one thing you need to know about this house is that the whole area as you go up on this side is a safe area. So, you can see that this will roll down." She points to a metal compartment above us, which neither my husband nor I had noticed. "I'm going to show you that all the hurricane shutters will also come down," she says.

The top of the stairs leads into the master suite. It is carpeted in ruby with burgundy walls. Gold curtains hang over the bed, which sits on a raised platform in the center of the room. The wall behind the bed is papered in gold filigree.

One corner of the room opens onto a green and white marble bathroom with a wall of mirrors. In the next room, we find two bookshelves and a leather-topped desk.

“The master study,” says my husband.

I kneel to peer through the double-sided fireplace and see the master suite on the other side.

“I think it does get chilly here in winter,” says the Realtor, “and so I think it’s neat to have a double-sided fireplace.”

“Forty at night,” says my husband in agreement. “Sixty, maybe sixty-five the next day.”

I leave the room. On the other side of the hall, I find a girl’s bedroom. The walls are lavender and pink with white crown molding. There’s a bay window to my left and a set of white wooden bookshelves piled high with worn books. A pair of autographed pointe shoes hangs on the wall before me.

“We have a dancer here,” I say, noticing the Realtor beside me.

“Are you a dancer?” she asks.

“I danced for a few years.”

“She’s being modest,” says my husband.

The bay window overlooks the length of the roof, which stretches on for a long, long time. I imagine going out there to be alone.

“She had a great room,” says the Realtor.

“How many kids were there?”

“Six. It’s a great house for a lot of kids.”

“You could easily sneak out this window,” I say.

“Oh, yeah.” The Realtor laughs. “So, all those hurricane shutters will come down. And those will come down.” She points to the bay window. “And those will come down, too.” She points to the other window.

It is rare to see poverty mentioned in Amway’s literature. When it is, it’s usually in the context of an Amway distributor having escaped it. Success is equated with wealth.[‡‡] With wealth is promised an enhanced way of life, one crafted of your own dreams—and Amway gives you The Plan to achieve that life. To let your attention stray from The Plan is to invite doubt and negative thinking, which can only result in failure. "As successful distributors tell people they are recruiting, the pursuit of excellence can be achieved only when they discipline themselves to tune in the positive dialogues and tune out the negative ones," says Cross. Poverty makes us feel bad. Feeling bad is negative. Negativity causes failure. It makes poverty feel contagious. So don't think about it.

Occasionally, though, it can be useful to mention poverty in a certain context. Inspired by the personal and business philosophies of DeVos and Van Andel, Cross spent the ten years after writing *Commitment to Excellence* researching the two men, culminating in his 1995 self-help book *Choices with Clout: How to Make Things Happen—by Making the Right Decisions Every Day of Your Life.* Much of the book is compiled from interviews with the Amway founders and top-level distributors. In a passage about excellence, Van Andel outlines the proper way for an Amway distributor to rationalize the issue of poverty:

> People think in terms of excellence, including success, wealth achievements, and gracious living. We feel uncomfortable about things at the lower end of the scale. We become anxious about peoples and nations in the grip of poverty. It makes us uneasy and often guilty to think of starving children and realize what bounties we have in America. Yet we should always bear in mind that poor

---

[‡‡] "Money—and what it can buy—is the universally recognizable indicator of success," says Cross.

> people cannot help poor people. What we can do, however, is to condition ourselves to speak out and stand up for those things in which we believe. To do this effectively, we must first have faith—faith in self, faith in God, faith in our convictions. Once these conditions are met, you will be amazed at how easy it is to speak out.
>
> Success usually requires sacrifice. You have to give up certain pleasures in order to devote time and effort to goals with higher priorities. Are you ready now?

Implicit in this is that those who have failed have failed because they have not made the necessary sacrifices to succeed. These are poor people. You're not a poor person, are you?

To achieve success through Amway, we must not only work hard but also have faith. We know that we should have faith in ourselves—Amway tells us this all the time. And we must have faith in our convictions—for instance, in the efficacy of free enterprise. The theologian, author, and "longtime friend of Amway and believer in its work ethic" Dr. Robert Schuller takes this one step further. In his writing he actually provides a list of six "existing strengths" in which Amway distributors should have faith, both individually and collectively: yourself, family, community, free enterprise, America, and faith itself.

But having faith in God, as Van Andel instructs above, is different. If you want to achieve success so that you can stand up for what you believe in, he says—you must have faith in God.

Amway's official position is that it doesn't endorse a single religion. Anyone can be an Amway distributor, it says—just look at the diversity of people around the world who are doing it. You see people of all races, political affiliations, and creeds succeeding in Amway.

And yet God frequently makes appearances in Amway's literature and teachings. Faith in God is given as the motivation underwriting everything from recruiting more distributors to persisting in The Business through every hardship to succeeding financially. Free enterprise is a blessing from God, says Amway. In *Simply Rich*, DeVos says:

> And then of course one question always comes up: "Should I even have this much wealth in the first place?"[§§] I feel the Lord allocated some money for us to use for our pleasure, some for our ability to experience His world, some for investing to help create economic expansion and job opportunities for others—and of course, some for sharing with those who have a real need.
>
> It's not because we're better or entitled to more money; we have been entrusted with it, and therefore need to be especially responsible. We just make sure personal spending doesn't become a priority over the giving side. Once you learn the budgeting process of setting aside for giving *first*, then what you have left you can allocate elsewhere—including a home or an airplane or a boat. One could always argue that these things aren't necessary and that you could give away more, and that's always true. But if you look at it that way, you'd never do anything more than take the bus.

THE BAYOU CLUB

7979 BAYOU CLUB BOULEVARD, LARGO, FL 33777

We've gone Diamond, we tell the membership director. We're joining the Bayou Club. We're starting a family.

[§§] In 2016, *Forbes* estimated Rich DeVos's net worth to be $4.8 billion.

The club recently underwent a $1 million renovation: new roof, redecorated dining hall and casual-attire bar and grille, revamped golf shop, locker rooms, fitness center, renovated driving range and greens. It closed for an extended period of time over the summer so that they could replace the greens and restore them to their original Tom Fazio PGA Tour–quality design. They use only Champion Dwarf Bermudagrass because, as the turf farm's website says, "even among the ultradwarf cultivars, there is no other grass capable of producing the incredible ball roll of a well-maintained Champion green."

We meet Dale at the top of the stairs at the Bayou Club's entrance. The room is cavernous—much bigger than it appears from the beach-colored exterior, where it seems miniaturized alongside the vastness of the course. He wears khaki pants, a too-small blue button-down shirt, and penny loafers. He reminds me of a shoe salesman.

He tells us the club no longer has an initiation fee—they were forced to waive it six years ago in response to the economic downturn. "You have the top two or three clubs in the area—Bayou Club, Belleair Country Club, and probably Feather Sound—with no initiation fees to join," he says. "It makes it very easy to be part of a club these days."

My husband seems pleased by this. His shoulders relax, and his gaze begins to wander around the place with familiarity.

We follow Dale into the casual grillroom. There are four square wooden tables and blue upholstered wooden chairs along a wooden bar that wraps around to another room with more tables. Floor-to-ceiling windows display the ninth and eighteenth holes.

"Shorts are fine here, jeans are fine. Casual attire, golf attire, tennis," says Dale. "What we train our staff on here, constantly, is the difference between a country club and a normal restaurant.

We have a membership: they're paying X amount of dollars just to walk in the door and come have a hamburger. So, we encourage the staff to make introductions if there are two members sitting here and they don't know each other. To get them involved, help them meet each other, help them make friends—because that's what's going to make them participate more and stay members longer. It's like a church. Like trying to get your congregation active and engaged and involved."

In the formal dining area, the windows are even taller, and peaked at the top. The area is bordered with potted palms and lit overhead by three chandeliers. Each table has a tall, skinny vase with lavender in its center. "We do a lot of weddings now because that's good business for the club," Dale says.

"This is a beautiful place for a wedding reception," I say.

"It is, and it's reasonable. I think that's where a lot of private clubs have leaned."

"Are those baby ducks out there on the course?" my husband says, moving toward the windows.

"A lot of wildlife," says Dale. "If we take the cart out, I can show you some of the real natural bayous in the golf course—the golf course is built around them to showcase that natural bayou setting."

We exit onto the back patio. It overlooks the pool on the lower level and the golf course, bordered by houses of the Bayou Club community. It's begun to rain lightly, but the sun is still out. We pass through an outdoor dining area and reenter through the fitness center: a room the size of a small apartment with mirrored walls, two rows of exercise machines, and a flat-screen TV mounted in the corner. A man and a woman exercise separately.

"We're just taking a quick peek," says Dale. "We'll be out of here in a minute. Morning, Dave!"

"Morning," says Dave.

"The gym used to be the men's smoking lounge," Dale says to us. "This is a pretty young club, but already we've seen a lot of changes. It's not all about the men saying, 'I want to join a golf club.' Now, with women having a much larger role in the family, they want to know, 'Well, what's in it for me?' There's got to be a fitness center, there's got to be some activities for ladies and kids, and it has to be more of a family culture. A lot of traditional men's golf clubs have had to really evolve into family clubs."

"Have you seen a racial shift in the membership as well?" says my husband.

"Not here in Florida. Not really. Pretty white around here. That's just how it works out.

"But I would say we're a pretty open club," Dale continues. "What we do have are a lot of same-sex couples at the club. That's become very common here. I think it's pretty popular in Florida and Tampa Bay."

On our way back to the front, Dale walks us through the pro shop. Two men lean across the counter talking to each other among displays of golf attire and equipment. One of them, the general manager, introduces himself to my husband.

"As long as you're a golf member, you're open to playing all the tournaments and games," Dale says to me. "There's something for the ladies, and then if couples play together, we have a couples' golf on Sundays. We have a senior group, and then a young under-forty-year-old guy group." He shows me a schedule pinned to a corkboard near the door. "These are kind of the core golf groups. And then we have a formal Men's Golf Association as well, one tournament per month. If they win that tournament, there are parking spots up for grabs, if you want a nice parking spot—or some trophies. You know, when you love a game and you watch it on TV, to be able to still play it and go out there with a large group of guys, and then win a tournament? These guys are having a blast. They feel like they're on the PGA Tour. That's what it's all about."

The general manager is making recommendations for the best seafood restaurants in the area. I tell him my parents live in Belleair.

"If you can't find good seafood around there, something's wrong with you," he says.

On the way out, we pass a frame on the wall bearing a quote by Robert Dedman Sr., founder of ClubCorp. My husband stops to read it: "'A club is a haven of refuge and accord in a world torn by strife and discord. A club is a place where kindred spirits gather to have fun and make friends. A club is a place of courtesy, good breeding, and good manners. A club is a place expressly for camaraderie, merriment, goodwill, and good cheer. A club humbles the mighty, draws out the timid, and casts out the sorehead. A club is one of the noblest inventions of mankind.'"

"The owners liked it so much, they called me up and said, 'I'm sending you something. I want you to put it on the wall,'" Dale says as we cross the parking lot toward a row of golf carts. "By the way," he adds, "you should know that Florida clubs tend to be less stuffy than northern clubs."

In 2015, *Forbes* named the DeVos family twentieth on their list of America's 50 Top Givers, with lifetime charity donations of $1.2 billion. Most of that money has stayed in West Michigan—Amway's headquarters are in Ada, and the DeVos and Van Andel families own or have bequeathed a considerable portion of Grand Rapids,[¶¶] and are of-

---

[¶¶] DeVos and Van Andel's first acquisition of property was the Amway Grand Plaza Hotel in 1979.

ten credited for catalyzing the revitalization of downtown. Of the $94 million the DeVos family gave in 2014 alone, $54 million of it stayed in Grand Rapids. Much of it went to public schools and Grand Rapids–based hospitals, arts programs, and faith-based organizations providing services to the homeless.

That same year over $4 million of DeVos's money went to Hope College, a private liberal arts school affiliated with the Reformed Church in America—in which Rich DeVos was raised—while $2.2 million went to Calvin College, associated with the Christian Reformed Church in North America.*** Of the $90.9 million in philanthropic donations the DeVos family made in 2013, 13 percent went to churches and faith-based organizations: $7.5 million to the King's College, a Christian college in New York City; $6.8 million to the Grand Rapids Christian Schools; and $1.05 million to the Chicago-based Willow Creek Community Church, an evangelical megachurch. As DeVos puts it in *Simply Rich*, "My Christian faith and outreach . . . remain strong after all these years. The Christian church and Christian education are high on our list of giving." He goes on to say:

> Collectively, our family has given away millions, but if the government increases our taxes by a big number, that makes it tough to have that number to give away. If the government takes it, then I can't give it—and I enjoy giving. *My* giving it puts the money in better hands than the government's.

*** DeVos has made it his life's mission to reunite the Reformed Church in America with the Christian Reformed Church in North America, from which it split in the 1850s, dividing his grandparents.

Each year, Rich DeVos attends The Gathering, a below-the-radar conference of hard-right Christian organizations and their biggest funders. Featured speakers have included the president and CEO of Alliance Defending Freedom, the president of Focus on the Family, and the head of the Family Research Council. The philanthropists in attendance are representatives of some of America's wealthiest dynasties and family foundations, and of the National Christian Foundation, America's largest provider of donor-advised funds given to Christian causes. Donors who meet at The Gathering dispense upwards of $1 billion a year in grants.

The family is also heavily invested in right-wing politics, earning comparisons to the Kochs for the enthusiasm with which they back Republican candidates like Newt Gingrich, Rick Santorum, Jeb Bush, Scott Walker, and Marco Rubio, and their sizable donations to ultraconservative organizations like Focus on the Family and the Family Research Council, both of which promote Christian value-based public policy such as anti-abortion legislation and bans on same-sex marriage. In 2014, the DeVoses donated in the six figures to Michigan-based conservative think tanks including the Acton Institute for the Study of Religion and Liberty, which promotes free market economics within a Christian framework, and the Mackinac Center for Public Policy, also a supporter of free market economics. Elsewhere, conservative organizations that received DeVos funding of over a million dollars each include the American Enterprise Institute, another free market think tank; the Alliance Defending Freedom, the right's preeminent legal defense fund; and the Heritage Foundation, which promotes free market economics and "traditional American values."

In the last quarter of 2015, DeVos family donations accounted for over half of those made to the Michigan Republican Party. Dick DeVos, Rich's oldest son, who served as president of the company before passing the torch to his younger brother Doug, made an unsuccessful run for Michigan governor in 2006. His wife, Betsy, has served as chair of the Michigan Republican Party and finance chair for the National Republican Senatorial Committee, and now chairs the board of directors of the American Federation for Children, a nonprofit which promotes giving students taxpayer-funded vouchers to attend private schools.

On a more personal note, Rich DeVos was close friends with Gerald Ford.[†††] They met when Ford was still a US congressman, and he regularly attended product launches when the company was still doing them out of DeVos's basement. As far as US presidents go, DeVos was also partial to Ronald Reagan—who appointed DeVos as finance chairman of the Republican National Committee and to the AIDS commission, about which DeVos has said:

> When HIV first came out, President Reagan formed a commission and I was honored to be on that commission. I listened to 300 witnesses tell us that it was everybody else's fault but their own. Nothing to do with their conduct, just that the government didn't fix this disease. At the end of that I put in the document—it was the conclusion document from the commission—that actions have consequences and you are responsible for yours. AIDS is a disease people gain because of their actions. It wasn't like cancer. We all made the exceptions for how you got it, by accident, that was all solved a long time ago.
>
> That's when they started hanging me in effigy because I wasn't sympathetic to all their requests for special treatment. Be-

---

††† In 2013, the family donated almost $1.4 million to the Gerald R. Ford Foundation.

> cause at that time it was always someone else's fault. I said, you are responsible for your actions, too, you know. Conduct yourself properly, which is a pretty solid Christian principle.

In *Simply Rich*, DeVos describes buying full-page advertisements for Reagan in popular magazines during his presidential runs because "we wanted the Amway distributors and their customers to know that we supported Reagan, in the hope that they would support him, too." Adding, "We also thought the ads might further help Amway distributors recognize the importance of free enterprise to their success." This is not the only time Amway has encouraged its sales force to back its political agenda. In 1994, Amway Crown Ambassador and motivational mogul Dexter Yager used Amway's extensive voice mail system to raise almost half of Amway distributor and "strong conservative" congresswoman Sue Myrick's campaign funds when she ran for North Carolina's ninth congressional district. The year Myrick was elected, Amway donated $1.3 million to the San Diego Convention and Visitors Bureau to pay for Republican "infomercials" airing on televangelist Pat Robertson's Family Channel during the party's August convention.

Yager, for his part, knows something about influence. He is nicknamed the "Master Dream Builder" and commonly referred to as a "hero" of The Business. Like most of Amway's top earners, he has been in it since its early days—Yager's business now employs four million Amway team members in over forty countries.

Yager made a name for himself as the father of the "Yager System," one of the first and most profitable motivational "tools" businesses run by Amway distributors (also called "tools scams" by detractors). Distributors produce motivational tapes and videos, or "tools," and sell them directly to their downlines for immediate profit. Tools promote Amway's free market philosophy but are not themselves Amway products—though the Yager Group is still today an Amway-approved training provider. The *Charlotte Observer* has said of Yager, "He sells not only soap but an ideology and a way of life. Admirers speak of him with reverence, as if his next plateau of Amway achievement were sainthood itself." The title of Yager's first book, *Don't Let Anybody Steal Your Dream*, was a Gerard household motto. We said it to one another with a near-religious zeal—like we were speaking in high-fives. I still feel nostalgic for my childhood when I hear it.

My husband rides in the front of the golf cart with Dale; I ride in the back. We strike out over the gently rolling fairways. "We're a longer course," says Dale. "Total length, if you play from back tees, seventy-one hundred yards. No one, not even the younger guys, play from the tips. I'm just going to show you the prettiest part and then head back so we stay dry."

We follow the right edge of the course, past houses hiding behind rows of palms: pool screens and burnt-orange rooftops flash by, one after another. Dale tells us that the country club owner's philosophy is not to overseed the fairways and greens but to preserve their natural beauty through proper maintenance. The tee boxes are overseeded with rye grass because people are taking strokes off them every day.

"We got some diehards here!" he says, slowing next to two older men preparing to tee off. They exchange hellos.

"He's a Clearwater fire chief," Dale says as we pull away. "And he's a pilot—a New Yorker."

A third of Bayou Club's membership is seasonal. People join the club to make friends because it's difficult finding a community when you're not from the area. Here, they know, people will remember their name.

A small waterway intersects the course ahead of us. We pull up to a wooden bridge, and Dale stops.

"We've got a little bit of surge here," he says. "Water levels are high. This is the Bayou Crossing Waterway. That way would take you out to Boca Ciega Bay, and eventually the Gulf of Mexico. When there's a huge tidal surge, these live bodies of water, the Bayou Crossing Waterway, feeds into, and overflows into, all these lakes and bayous around the course. And then when the water recedes, any fish and the water that gets in there gets trapped in there and can't get out."

I snap a photo with my phone as we cross the bridge.

"Tom Fazio is the golf course designer," says Dale. "He was pretty green and environmental-friendly, kind of before it was all really cool, and that's what he does, is he builds the course around the natural landscape without changing it."

"Doesn't the golf course itself change it?" I say.

"Does the course change the landscape?" says Dale.

"Yeah, doesn't the golf course itself change the landscape," I say.

"It does, but I think he tries to not change the natural landscape, but to design the course around it," says Dale. "Like, all these holes going right next to the Bayou Crossing Waterway. It's . . . it's . . . it's part of it."

We drive on.

"And then the bayous, the natural bayous, having those as, um, hazards to hit over and around," he says. "It's just fun."

"I bet there are some nice balls in these bayous," says my husband.

"You know, we actually have a guy who's on contract with us. He pays us an annual fee to remove the balls from our lakes and ponds."

"What kinds of species do you see out here?" I say.

"A lot of mullet, some sea trout," says Dale. "Birds, some blue heron, some osprey, ibis. Definitely some gators. We don't reach too far into the bushes to get our balls."

Dale hangs a left, and we circle back toward the clubhouse. Halfway there, he slows by a pond. A family of ducks crosses the course and alights on the water.

"What happens if we're members for six or seven years and suddenly we can't afford the membership fee anymore?" says my husband.

"Then you can resign," says Dale. "We have a medical leave program if there's a medical issue. If it's a money thing, it's financial. We do lose members, but that's not something either you or we can control."

Back in the clubhouse, Dale shows us the locker rooms and the administrative offices, pointing out the one for the Bayou Club community association. He pauses before a plaque outside his office, displaying several rows of names.

"One of our traditions is this Hole in One Club," he says. "We don't use this plaque anymore, but we do make a plaque with a picture of the hole and the date you made it and your name. Some people go their whole lives and never make a hole in one, so we make a big deal out of it. You have to have a witness—you come back to the clubhouse, your witness has to verify with the pro shop. Then we open a free bar tab for you for the rest of the day. All golf members are part of it, so the insurance on it is: If someone makes a hole in one, every golf member is charged one dollar. So, that creates a three-hundred-thirty-dollar credit that you will receive. If you don't use it at the bar, you'll get a certificate to use around the club for anything else."

"I've always wanted to make a hole in one," says my husband.

"It's a rare thing," says Dale. "Sure, there's a lot of skill—you have to hit it just right—but in the end, it's kind of luck. It's kind of a little bit of luck, too. Maybe eighty percent skill, twenty percent luck."

"Have you ever gotten one?" says my husband.

"No," says Dale. "I've only been playing seriously for six or seven years, and I don't have much time, working in hospitality. But I love playing at Bayou Club. You join a private club hoping that during season when every other golf course is swamped—I mean, we own a public course nearby, and they're running on six-minute tee times. They're herded through there like cattle. It's tough during season, and it's not enjoyable golf. Because if you're playing golf, especially if you're kind of a quick player, when you run into someone else and then you have to stop and you have to wait for those people to play ahead of you, to get out of the way, it interrupts your rhythm playing the game."

He walks us to the front with a membership application. We promise we'll return it to him within the week—then walk across the parking lot back toward our Porsche.

My parents more or less broke even in Amway. They didn't lose any money; they also didn't make any. I learned recently that my mom was against it from the start. "She never wanted to do it, never warmed up to it," says my dad. She believed it was a cult, and wasn't happy about giving their time and money to it. She hated Amway's right-wing political propaganda and evangelical bullying. She hated that it kept the two of them from spending time with me. "She wasn't going to leave me," my dad says. "But there was tension because she didn't want to go do these things." Even as he admits he agreed with her on some level, he wanted to believe that The Business was viable.

In four years, they built up their downline to something like forty people. It was a cumbersome organization, but the people they were working with, save for one, were all honest. A lot of them had families we'd grown close to—the kids were my friends. I'd go to their houses on the weekends, and after school, and whenever my parents needed a babysitter. After we left Amway, I never saw them again.

In the beginning, my parents put between ten and fifteen hours a week into their business—per the company's recommendation. But over time, my dad's enthusiasm began to wear off. "You say to yourself, 'What the hell for?'" he says now. "So that somebody can come in and then not return your calls? You take them to a meeting and there's a jerk up there who's embarrassing? I had no way, no avenue to get people in there and get them excited."

The embarrassing jerk was my parents' upline, Vincent, who had Emerald status. I don't remember this man. My dad says, "He was a creepy guy, just an incredibly creepy guy. I don't know how else to describe him . . . You actually felt, after being around the guy, that you needed to take a shower. Nobody wanted to be around him. He was a jerk, he was a liar. Just a despicable person."

This is not the man who brought my dad in but a man somewhere above him. He was what The Business calls a "phony Emerald." To meet the criteria for the pin level, he'd force the people in his organization to order extra product in order to grow his volume and push him across the finish line each month—not that he turned much of a profit doing so, as he had to pass it all on to his own upline. "Well, the Emerald pin doesn't mean anything unless your organization is solid," said my dad. "So you got a pin—you're not making the money." Eventually, my dad says, Vincent was stripped of the Emerald pin because he couldn't maintain the sales by force alone.

It was hard enough to get people to sign up for Amway. My parents, in describing their experience, said that most people had heard of the company and believed it was a pyramid scheme. In fact, part of my parents' strategy for "showing The Plan" was that they didn't even tell people it was Amway until the very end of their presentation—then they signed them up on the spot. If they couldn't sign them up right then, they invited them to a meeting. Most of the time, even though they told them not to talk to anybody about Amway before the meeting, the prospect would go to their brother-in-law, who would tell them it was crap. "And if they make it to the meeting, this guy"—the creepy guy in the upline—"stands up there and is a complete ass," says my dad. "And the people that you encouraged and cajoled, they take a look at you and say, '*What?*' And then they don't return your phone call."

Not everybody in the world has a big vision of what life can be. Most people go to work, they get a check, they go home, and that's their life: they don't have a big vision.

"To build that in somebody is a great effort," says my dad. "And then, once you do build it, to have somebody go and tear it apart because he's a jackass . . ."

On top of that, my dad's advertising business was taking off. He didn't have time for two entrepreneurial ventures. Forced to choose between them, the best option was clear.

I think of my family's time in Amway as achievement tourism. We left reality for a moment and believed the impossible was possible. My dad still wonders if there's more he could have done, if there's a way for him to have succeeded in Amway—admitting in the next breath that there isn't. My parents tried everything. At each turn, the people they thought were supposed to be helping them—their upline, yes, but really the overall structure of the Amway Corporation itself—actually stood in their way. They built dreams and worked to achieve them, but the only people who benefited from their work were the people already on top.

For my part, I'm now skeptical of my materialist impulses. The dreams I built in Amway don't appeal to me anymore. I find them claustrophobic. Ultimately they made the walls close in on my family as we reduced our visions for ourselves to what we owned rather than who we were and how we lived our lives.

Directly across the state from my family, on Florida's Atlantic coast, is the Windsor country club. Home architecture here is strictly regulated. Residents drive around on golf carts, on and off the eighteen-hole course. There's an equestrian center, tennis courts, a concierge, and a gun club. Occasionally Prince Charles pays a visit. This is where you go when you bypass Palm Beach on your way to vacation—there's no kitsch in Windsor, only the highly refined. Among its residents are retail billionaire W. Galen Weston, the Swarovski clan—and the DeVoses, who own three houses here and spend eight weeks a year or more on the waterfront.

"I never wanted to have a house in Florida," Betsy DeVos explained to the *Wall Street Journal*. "But Windsor is so different from the rest of Florida."

There's no such thing as class in Windsor—everyone is as rich as everybody else. In Windsor, Rich DeVos can catch some rays in peace. No one bothers him about "ethical this" and "fraudulent that." He plays golf all day. He never has to mow the lawn or wait at a traffic light.

In Windsor, Rich is surrounded by civilized people. There are no termites. The pool is always eighty degrees. The beach is walking distance. There are no sharks in the water, even at night. Birds never shit on his car in Windsor. There are no loud tourists in Windsor. There's no media. There's plenty of shade. There are no alligators.

The people are all Rich's friends in Windsor. People always

agree with him here. In Windsor, there is only small talk. Everyone donates to the charities of Rich's choosing. He gets a hole in one every day.

In 2010, Amway reached a settlement reportedly valued at $100 million in a California class action lawsuit filed by three former distributors who claimed the company was operating as a pyramid scheme. In addition to paying the plaintiffs and their attorneys, the company announced in a letter to its employees that, as part of the settlement, it was taking action to address many of the concerns raised in the case. Among the actions taken were tripling investments in IBO education programs and more than doubling the number of professional trainers, such as the Yagers, across the country. A year after the California case was settled, Amway offices in India were raided for the second time among multiple complaints about the company's practices and its upper-level distributors. The following year, they were raided again, and the CEO of Amway India was arrested for fraud.

In 2014, Founders Crown Ambassadors Barry Chi and Holly Chen, who run the biggest Amway distributorship in the world based in Taiwan, were sued by nine Chinese immigrants in the Southern California region who claimed that, although Chi and Chen promised they could potentially make millions in commissions, Amway business owners make closer to $200 a month.

Some things never change.

Records

All names have been changed.

I meet Jerod my senior year of high school. It's the year I get my braces off, the year I switch from glasses to contacts. He's a senior, too, though we go to different schools: I'm in the arts magnet studying music; he's in the technical school across the street, where he's earning his GED. The day we meet, he's working at the Chick-fil-A in the Tyrone Square food court and charms me into giving him my phone number, then a cigarette. We smoke in the parking lot behind a row of Dumpsters, putrefying under the midday Florida sun. He has dark red hair, amber eyes, and full lips. His fingers make delicate little movements with his cigarette. He asks me what kind of music I like.

"Bright Eyes."

"Never heard of it," he says.

"Elliott Smith?"

He shakes his head. He tells me he's DJ Qwork and he spins electro. He has a set of Technics turntables in his room and milk crates full of vinyl. Once a month, he hauls his crates to Ybor

City and waits outside the Amphitheatre for the chance to spin, but mostly, he just spins for friends in his room, or at house parties, or at raves under the Sunshine Skyway. He always knows when raves are happening. His favorite thing to do, after spinning records, is to hold forth about the history of electronic music and the nuances of the subgenres: jungle, drum 'n' bass, house, ambient—there are hundreds. He has a tan hat with the word "electro" embroidered on the brim. He wears it every day, always perfectly clean.

We're seventeen, but he's older streetwise, has lived a whole life before me—so he thinks I'm sweet and I like proving that I'm not. The year before, his mother sent him to live with his father in St. Petersburg, to escape what she considered negative influences. The way he talks about his recent past in Tampa is thrilling to me. He's freebased cocaine, shot heroin, fucked a lot of girls—by this time, I've seen five penises and slept with three people, maybe a total of ten times. Shortly after we start dating, he graduates to the smoothie counter next door to the Chick-fil-A. It's also when he meets Sean, who becomes our main drug connect. This is where my year of living dangerously begins.

My high school is an arts magnet program designed to bring white kids into a black neighborhood. It's a school within a school: maybe four hundred kids in the magnet, out of two thousand total. Theoretically, anyone can enter the arts program, but you have to audition to get in. So, it helps if you also went to the arts elementary school and the arts middle school, which I did. Since fifth grade, I've studied ballet, character dance, jazz dance, vocal music, musical theater, and visual art. I also served a brief stint in the handbell ensemble.

In my sophomore year of high school, I changed my major from musical theater to vocal performance—fleeing musical theater's narcissism—and joined the choir. The choir is austere. We study opera. Once a week, I leave choir class to take private singing lessons with Mrs. Bancroft in a tiny studio with an upright piano. At the end of my senior year, I'll give a finale recital with my best friends, Gisele and Lily.

My other best friend, Ashley, is a dance major. We met at theater camp the summer after first grade. Ace of Base's "I Saw the Sign" was the hottest thing on the radio after Mariah Carey's "Dreamlover." We played Ace of Base on the camp stereo every morning, and I sang Mariah Carey's "Hero" at the end-of-the-summer show. I've always been a singer. But Ashley's always been a dancer. She started when she was six and has stuck with it all the way until now, our senior year, eleven years later. The dancers are known for being kind of slutty, and there's a rumor that some of them can smoke cigarettes with their vaginas. Ashley isn't a slut, but she embraces the bad-girl persona. She's lanky, and her weight fluctuates constantly depending on whether she's eating or not. She smokes a pack of cigarettes a day. She pulls off tiny shorts better than anyone else I know.

Our senior year, Ashley meets Miles. Miles works at the Petland next to the movie theater in Largo Mall. We hang out at the movie theater whether or not we're seeing a movie. Ashley recently started a job there, so now we're friends with all of the employees. At the theater—in general—we collect friends. Not friends with common interests, but friends whose common ground is an overall lack of interests. They're simply around, gathering in empty spaces like dirt swept into sidewalk cracks. Miles is one of these people. He appears to know a lot about animals. He keeps a menagerie of reptiles in tanks next to his bongs. He's seven years our senior, and he really likes pills.

Miles seems harmless at first. He makes eye contact. He's okay-looking: sandy hair, green eyes that are a little too close together. And he knows that we crave male attention, so he flirts with us, which we think is fun. He gets jealous, which Ashley finds flattering, at least in the beginning. He always has weed, and sometimes something extra—Xanax, Vicodin, ecstasy. Ecstasy is the new designer drug, and everyone who does it thinks he's an expert on it. Miles rattles off the formulas of pills he buys as though he pressed them himself.

Ashley's mom has cancer, and she's missing a good chunk of our senior year staying home to take care of her—or saying she's home. It's also hard for Ashley to be in the house where her mother is dying, so she finds ways not to be there. Miles becomes a good reason not to be there. Soon after Ashley starts dating him, she tries ecstasy and discovers its particular breed of euphoric joy. She starts doing it all the time.

Miles moves around a lot. When we meet him, he's living in a powder blue duplex off of Ulmerton Road, in an unincorporated area between Largo and Pinellas Park considered trashy. Some of our friends from the movie theater live there, and within a few months of meeting Miles, Ashley is basically living there, too. She sleeps at Miles's and shows up to school at seven thirty the next morning hung over, still in dance clothes. After school, she drives home to check on her mom and then goes to the movie theater, where she works for a few hours, and then to Miles's, where she spends the night. Her own house becomes a place to change her clothes. She comes to school less and less. Then she stops coming altogether.

I don't really know what goes on at the duplex. Concerned, I show up there one Saturday afternoon. Miles is sleeping. Ashley and I sit on an overstuffed sofa with Miles's friend, a metalhead, thirtysomething guy named Clark, who works at an electronics store. We smoke cigarettes with the windows closed and the blinds drawn. At some point, Ashley tells us Miles is sleeping off a Xanax high. She says, "I'm going to go wake him up with a blow job."

I sit alone with Clark on the sofa. I don't know him well, but I've heard his girlfriend is scary. I saw her once; I remember little more than the heart-shaped padlock tattoo that covered her whole chest. Clark and I make small talk. Within ten minutes, Ashley is back. She's crying. She marches into the kitchen, turns on the stove, and starts to silently make an egg sandwich. Clark and I follow her with our eyes. Midway through, she returns to the bedroom, and we hear shouting. She's back out a minute later, slamming the door. We ask her what's wrong.

"He's a fucking asshole when he wakes up from Xanax," she says.

She finishes the sandwich and takes it in to him. There's more yelling. I leave.

With Jerod, smoking weed is a way of keeping time. It's a ritual that marks the end of one state of being and the start of another: separateness, togetherness; stillness, goingness. When I arrive at his house after school, we smoke a bowl. When Sean comes to pick us up to go to a party, we smoke a bowl. It welcomes new people into the fold. Seeking out more of it gives us a mission. It mobilizes us so that we can make ourselves once again immobile, and celebrate our weed-seeking victories by smoking more weed while watching cartoons.

Stoned in his room, we talk about music. We agree on Kraftwerk and certain Pink Floyd albums. We disagree on almost everything else, but we tolerate each other's tastes. I come to like Anthony Rother, and Jerod comes to like Radiohead. I'm in my third year of music theory classes, and I like to count out the time signatures for him, lying on his bunk bed while he spins. He likes the grinding bass and driving drumbeat of electro. It's dark and beautiful. It makes him feel fierce. When he spins, he presses his headphones to one ear and bobs over the mixing board, wiggling his fingers. When he gets to his favorite parts, he lifts his hands into the air like he's praying.

He knows what he wants and who he is. He has a bright future. It's glow-lit, a laser light show. His favorite movie is *Groove*; his favorite book is *Last Night a DJ Saved My Life*; his money goes toward drugs and records. He isn't complicated. He's determined.

He explains his DJ name to me. "A Qwork is an inherent quandary," he says. We're sitting on his bedroom carpet after a make-out session. Rave posters hang on his walls among glow-in-the-dark stars and comets. He reaches over to his backpack and retrieves a spiral notebook and a pencil. On the first empty page he draws a wiggly half circle. With the two loose ends, he completes a square, then draws another square a little ways off and connects the corners to make the image look three-dimensional.

"It's a cube and a sphere smashed together into one entity," he says earnestly, looking me in the eye. "It's self-made."

We roll every few weeks. Sometimes Sean scores the ecstasy, sometimes another friend. It's a warm October when Sean picks up Jerod and me and drives us to a party in one of the beach neighborhoods. I don't know the person throwing it; we meet at a different house each time. Jerod is supposed to spin at this one, so he's brought his records, which ride in the trunk. On the way, we pick up Gisele. She and I have been best friends since freshmen orientation. It's her first time rolling—she's nervous about it and wants to be around people she trusts.

Gisele is petite and half Yemeni: curly black hair down her back, cardamom skin, sharp brown eyes, and an asymmetrical smile. She's gaga for Cirque du Soleil and has aspirations of being a performer—she's trained herself in hand balancing and is a virtuoso with glow sticks: a favorite at every party. If hand balancing fails, she plans to fall back on acrobatics. If that fails, she'll be a singer. She's one of the best singers in our class, one of the best I've ever heard, and already has inside connections at Cirque. She's made friends with a hand balancer from the show *Quidam*, and with a costume artist from *Dralion*. She's spent a week in France with the costume artist and her ex-boyfriend, a vocalist from *Quidam*. She knows all of the Cirque songs by heart. When we graduate, she plans to go traveling as an usher with the show *Alegria*, with hopes of apprenticing under the vocalists.

It's magic hour when we arrive at the party. The house sits on stilts with a screened-in balcony facing the street. A shirtless man lies on the floor of the living room as we enter. Others huddle in various corners. A set of turntables waits against the wall by the sliding glass doors that lead out to the balcony. Within an hour, we're grinding our teeth and chain-smoking Newports, and blowing Vicks inhalers into each other's eyes. The balcony is the most beautiful place I've ever seen, and everyone on it is part soul mate, part genius. Gisele and I kiss to solidify our connection.

TJ appears next to us. TJ dated our friend Lily in middle school and took her virginity. It's generally understood that he did her wrong somehow, but none of us knows exactly what happened. His head is shaved from the bottom of his hairline up to the tops of his ears. The rest is dreadlocked and pulled into a bundle. He wears a white T-shirt, parachute pants, and skateboarding shoes. We've seen his picture before but never met him. Before we know it, we're deep in conversation.

"Spinning is an out-of-body experience," he tells us. He's introduced himself as DJ Mission, one of the DJs spinning tonight. His specialties are trance and techno. He's just finished and DJ Qwork is on.

"We're singers," Gisele says. "For us, it's not out-of-body, it's *from*-the-body."

"An in-body experience," he says.

"*Coming-out-of*," I correct him.

"Your spirit touches the audience," Gisele says.

Sometime later, they're gone and I'm melting ice on a stranger's chest. I recognize him as the man who was lying on the floor when I first arrived. He's still on the floor. His eyes are closed and his hands are up above his head. His chest is tan and muscular. Glow sticks reflect in the rivulets of water trickling over it. He's telling me about his job as a public speaker.

"I'm the oldest living person born with AIDS," he yells over the music. DJ Qwork is still spinning. I wonder how long I've been doing this. I really like how it feels. "I go around to high schools and old people's homes and I talk about overcoming hardship."

"What do you tell them?" I say.

Gisele and TJ return. We're again on the balcony. Jerod is with us now. His eyes are closed and I'm rubbing his cheekbones. A bottle of Vicks VapoRub appears next to me. I wipe some under his nose. Gisele and TJ are holding hands, and we all accept it; we want them to be in love; love is beautiful. I light a cigarette. I light another off another. I make out with Jerod and then go back to making out with Gisele. I take my shirt off and empty a bottle of water over my head and lie on the floor. Gisele is in the living room spinning glow sticks, and we're all in the back of TJ's car riding home at sunrise.

VOCALISTS ARE REQUIRED TO PERFORM AT THEIR SENIOR RECITAL TWO ITALIAN SONGS, TWO French, two German, and two English. We sing Mozart, Bach,

Handel, Schumann, Strauss, and Berlioz. We sing Irish and American folk songs. We sing jazz. Lily's best aria is Mozart's "Batti, batti, o bel Massetto" from *Don Giovanni.* It's passionate and heartrending; it fits her personality. Mine is Caccini's moody madrigal, "Amarilli, mia bella," full of torture and vibrato. Students' private voice instructors assign six of the songs, and the other two can be songs of a student's choosing. I plan to sing, as the songs of my choosing, Handel's aria "Lascia ch'io pianga" from *Rinaldo* and "Sally's Song" from *The Nightmare Before Christmas*—Gisele will be performing "Jack's Lament."

I'm looking ahead to what kind of artist I want to be once I leave the protective confines of the high school. Though I like singing opera, I'm not passionate about it. Singing other people's songs for a living doesn't appeal to me. I'd rather sing my own. My training is in choral music, so I explore composing for choir: I take a sheet of staff paper from my music theory notebook and begin to compose an "O Magnum Mysterium," mostly because I like the one we'll be singing in the winter concert. I draw the treble and bass clefs and ponder the first note. All I know about composing is what I've learned in music theory class. I don't play an instrument. I know little about the tradition of the form in which I'm writing. I suddenly feel like I'm attempting to write a novel knowing only the basics of English grammar. I don't make it past the first line.

I've been taking guitar lessons with my uncle and toy with the idea of pursuing a path as a singer-songwriter. I even cobble together a few songs. My guitar is an acoustic Austin I bought at Sam Ash for $300. I call it Betty—it's definitely a lady. I like to tune Betty; tuning is something I feel I can do well. After tuning her, I like to play a G chord and then move my ring finger from the lowest string to the string above. I'm not sure what this does technically, but I feel it improves the overall sound of the chord, so I implement this trick in all of my songs. Simple and depressing, my songs employ as many combinations of E, G, D, and C chords as I can muster.

I enjoy the singing part of being a singer-songwriter. Since childhood, I've trained myself in a soprano register, though I'm more comfortable singing alto. I feel that sopranos are, de facto, truer vocalists. I have a clear, high voice, and can perform complicated variations in pitch. More than any other kind of song, I enjoy ballads—the musical theater instructor once told me I sound like a country singer. But more than singing, I enjoy writing lyrics. I feel the poetry of my lyrics and the quality of my singing voice together outshine my limited proficiency with guitar. Sometimes my lyrics evolve from poems; sometimes I begin with the intention of writing lyrics and then just continue writing when I reach the end of a song. My journal entries are sprawling and emotionally wrought. A few times, I read them aloud to my parents—to communicate something otherwise ineffable? I don't know. I consider my future persona to be an admixture of Ani DiFranco, Tori Amos, and Sylvia Plath. I have a lot of complicated feelings and someday people will care about them.

This semester, I'm also enrolled in a photography class. Each student in the class has been assigned a camera, and I carry mine everywhere. I love feeling the weight of it around my neck. I imagine invisible rectangles framing everything. Being in the darkroom is a way to obliterate time. I love the magic of the image emerging from the white, and the smell of the chemicals. The word "aperture" sounds like kissing. "Shutter" sounds like butterflies. I begin telling my friends of my plans to pursue a path as a photojournalist—specifically in the world of rock music. I'll be a staff photographer for *Rolling Stone*, I say. Traveling with my favorite bands will be my job. I'm going to see things I can never see otherwise. Just think of all the people I'll meet.

My parents and I go to New York City to visit colleges. We rent a room on the tenth floor of the Helmsley Hotel on Forty-Second Street. From the nearest corner, we can look west and see the lights of Times Square. It's my first time here, and it's night. My parents have gone to bed, and they've specifically told me not to leave the hotel unaccompanied. I'm outside on the curb, pointing my camera west down Forty-Second Street, bringing the distant glow into focus.

"Whatcha lookin' at?" says a voice behind me.

I lower my camera and turn around. Two men in suits smile back at me. The taller one has Italian features: olive skin, thick hair, and a trimmed beard. The other is younger, lighter-skinned, with light eyes.

"Taking pictures," I say.

"Are you a photographer?" says the tall one.

"I'm just taking a class."

"Oh? A college class?"

"High school."

He smiles slyly. "I bet there's not even film in that camera."

I look down at my camera, then down at my outfit. It's chilly; I'm wearing a baggy faux-fur Nine West jacket and closed-toed Birkenstocks. I've drawn all over the thighs of my jeans.

"It has film in it," I say. I hold the camera out to show him, pointing to the window in the back that displays the roll.

"You know, we're photographers," he says.

"Really?"

"We are. We shoot for *National Geographic*."

"Wow," I say.

"How come you have your aperture stopped down so low?"

I look at him. I know this word "aperture" and can locate it on a camera, but I don't yet have a grasp of its function. When taking a picture, I simply turn the wheel on the top of the camera and the one on the front until the light meter is green. I can fix a good deal of my camera-related errors in the darkroom, or I feel that I can. The ones that I can't, I simply call "abstract."

"I don't really know," I confess. "We haven't really learned that in class, yet."

"Bullshit," he says. "You're not in high school."

"I am," I say. I take my wallet out of my pocket and show him my driver's license. In the photo, I'm wearing a burgundy Bad Kitty T-shirt, dangly earrings, and heavy eye makeup. "My birthday is right here," I say. "I'm seventeen. I'm here visiting colleges."

He looks at me and then back at his friend, who's been standing a few feet away from us while we talk. His friend smiles and shakes his head.

"Listen," says the tall one. "You don't want to be taking pictures of Times Square. Everyone takes pictures of Times Square—you're not going to take one better than they did. You need a story."

"What do you mean?" I say.

"I mean a story, something to care about, with people. Come here, I'll show you."

He puts his hand on my back and starts to guide me. We turn off of Forty-Second Street down a darker street that's mostly empty. We walk quickly and silently. Midway down the block, we stop outside of a bar.

"Here," he says. "Look at this."

A man and woman smoke outside, silhouetted by a streetlight.

"What about it?" I say.

"What about it? This is a story," he says. "They're arguing. There's conflict."

He's right—I notice it now. They aren't yelling, but their body language is tense. Still, I don't connect with the composition.

"Take it," he says. "Come on."

"What's interesting about it?" I say.

"Point the camera and take the picture. Trust me. Do it now."

We're in a restaurant around the corner from the bar. They sit on one side of the table, I sit on the other. I'm trapped by the wall behind me, the window to my right, and the empty chair to my left. They ask if I want to order something and I say no. When the younger man's coffee arrives, I say that I'm going back to my hotel.

"No, no, wait a minute," says the tall one. "Stay. We were just talking."

"I can't," I say. "I'm sorry, I have to go back now."

"Just stay for a minute."

"I really can't."

He looks at his friend. His friend shrugs. "Let her go," he says.

Outside on the curb, the tall one insists on hailing me a taxi. When one arrives, the younger man climbs inside and grabs my arm and yanks me downward. I yell at him to let go, pulling backward. A twenty-dollar bill appears in his hand, and before I know it, he's putting his hand in my jeans pocket, and stuffing the bill down inside. I jerk myself free, repeating the word "no." The taxi pulls away. I run the three blocks back to my hotel and go directly upstairs, my heart pounding.

My parents are awake when I come into the room. They're angry that they don't know where I've been. I lie. I tell them I was downstairs on the mezzanine floor, taking pictures from the Helmsley's balcony.

Back in Florida, I develop the picture I took outside the bar. I find it underexposed: the focus soft, with no details. The sign on the bar awning reads "Attitude Adjustment Hour." I feel afraid and ashamed looking at it, knowing that this is not a picture I wanted to take; that I took it only because I was told to.

I carry it around in my binder for weeks, wondering what to do with it, unable to let it go. Each time I remember it's there, I feel the man's hand again on my lower back: the way he guided me, the way I let him. I hear again his telling me that my photographs need a story. I think about this with obsessive concern each time I compress the shutter. I think about it as I write in my journal. I think about it as I play my guitar. I see the photograph in my binder each time I open it, feeling again the jolt of terror and the yank on my wrist. Over time, the jolt becomes a quick tug, then a vague sickness, then a faraway cringe.

I can't bring myself to tell others about that night. But the picture is nonetheless a record of it that I can't afford to lose. Finally, feeling I can't throw it away, I frame it and hang it on my bedroom wall.

Miles has the idea to sell Ashley's birth control placebo pills on the beach as ecstasy. Clearwater Beach, always packed, is their best bet. They go down to the boardwalk on a weekend, when the knickknack peddlers line up their booths on the pier and fire dancers spin by the water and college students get wasted and stumble down the strip. They repurpose the baggies their own beans came in and target dumb teenagers desperate to get high. Ashley is there to look trustworthy. Miles talks up the quality of the drug. They charge $25 a pill. They make $175.

We meet at the duplex on a Friday night. I've brought clothes for sleeping over: baby blue pajama pants that I put on immediately, my elementary school T-shirt, and bunny slippers. There are glow sticks waiting in bags on the table: someone purchased them at Spencer's Gifts earlier in the day. We've each brought extra packs of cigarettes and Vicks inhalers. Clark is there—by this time, I've decided he's cool. Miles turns off the lights in the living room and turns on the black light above the TV. We crush our beans on the coffee table and snort them through rolled-up dollar bills. Clark rolls a blunt to facilitate our highs while the ecstasy kicks in.

I'm sitting sideways on the couch, legs extended before me, my arms crossed over my chest. Ashley is sitting behind me with her arms wrapped around mine and her legs extended on either side of my own. "Close your eyes," she says, "and breathe with me."

We breathe in and out; in and out; in and out.

We fall backward together, and she pulls me down against her chest. My eyes roll into my head and flutter. We rock back and forth, back and forth. Warmth spreads from the top of my head into my groin and gathers there, trickles down to my feet. Ashley's hands find my face and massage my temples. Someone drips cool water over my forehead and drags it through my hair.

I'm speeding over a digital freeway, and beneath me is a curling baseline that arches into my spine. A man's voice traces the line of a mountain range in the distance and I feel it in my belly. I hear Miles then Ashley, but I don't know what they're saying, so I open my eyes. We're back in the same living room, on the same night as before, only everyone looks different. Dirty Vegas's "Days Go By" is on the speakers.

"Do you see his glasses?" Ashley says to me. She's cross-legged on the floor with a pillow on her lap. She's changed into a sports bra and basketball shorts. Her hair is gathered in a messy bun on top of her head.

"I swear I'm not wearing glasses," says Clark.

"You are!"

"I'm not."

Miles is sitting next to me, I realize. His bent knee is resting against my thigh. "I've heard that before," he says, shifting his weight so we're touching more. "It's the ecstasy. Some people see glasses."

"I see them," I say.

"You do?" says Ashley.

I look at Clark, who turns to face me. We gaze at each other. He's prettier than I remember.

"I see them," I say.

Each Wednesday after school, the music department holds student recitals, or "juries," in the small performance hall in Building 3. Juries are graded the same way as senior recitals, but on a smaller scale, and attendance is mandatory. Roughly ten students perform each week, from the instrumental and vocal music departments. Every music student has to perform three times a semester, one piece each time. Typically, these are pieces they're already practicing in private lessons. Juries terrify me.

It's just before the winter break. I'm lining up to perform Schumann's "Er, der Herrlichste von Allen," a song I hate. The ascending lines are aggressive, stampeding—like barging into someone's room uninvited. The German language to me is equivalent to barking—there's no poetry in it. I'd much rather sing Italian. Besides, the translation of the song is embarrassing: the speaker is a woman who's fawning over a man, showering him in the dorkiest praises: "O, how mild, so good!" It's nauseating. I inherited the song from Mrs. Pfister, my voice teacher the year before; she pressured me into learning it. Imagining singing it now, in earnest, in front of my classmates, and projecting from my diaphragm, fills me with dread.

I begin to cry. There are people behind and in front of me: Trang, a timpanist; Billy, a trombonist; Kelly, a vocalist; and others I know from the music department. I'm aware that as I cry, I'm producing mucus and my throat is swelling, both of which inhibit singing. This makes me cry harder. I count the number of people ahead of me in line: three. I take several deep belly breaths. Trang turns around to ask if I'm okay and then steps into the nearby bathroom to get toilet paper. I'm suddenly aware that I'm crying in front of Billy, which further embarrasses me—I've always had a crush on Billy. The thought of singing "Er, der Herrlichste von Allen" in front of him is the most mortifying thing I can think of.

Before I start to cry again, before I realize what's happening, I'm outside the building. The campus is empty; all of the zoned students left after sixth period and the art students left after seventh—only music students remain. The sun has been setting earlier each day as Christmas approaches, and the darkness of the late afternoon drops another layer of silence over the already silent campus. I squat on the sidewalk and hang my head between my shoulders. Rescheduling the recital is not an option. I procrastinated in signing up for my third performance, and now there are no slots left in the remaining recitals before the break. It occurs to me that this wasn't going to be a good performance, anyway: in general, I have an issue with "projecting"; my voice is not very loud. That's why Mrs. Pfister assigned "Er, der Herrlichste von Allen" in the first place—so I'd learn to sing louder.

I hum scales as I go back inside in a last-ditch attempt to soothe my voice after crying. I stop at the water fountain, aware that the icy water is not doing me any favors. Billy is performing; only Trang is ahead of me. I now see that Ms. Loup is on the piano. I've been taking private voice lessons with Ms. Loup since I was seven years old. We used to meet in her house in Clearwater on Saturdays. In her living room was an upright piano and stacks upon stacks of musical scores. It was there that I learned the name George Gershwin. Ms. Loup taught me "Lascia ch'io pianga"—one of the songs I'd chosen to sing in my senior recital—when I was twelve years old. It was coincidence that led her to work at the arts high school the same year I entered it. I feel instantly comforted with her here.

I clear mucus from my throat as I step into the hall. My vocal cords are sore, but there's something about it that I like—a battle wound. Ms. Loup pounds out the opening chords of "Er, der Herrlichste von Allen," and I make a conscious decision not to care about how my performance is received. I don't give a shit about opera, anyway: I'm going to be a singer-songwriter, or a music journalist, or a poet. I take a deep breath and open my mouth. I close my eyes.

I'm kissing Eric Steeler on the quad. It's after lunch; we've been sitting together recently, and today he decided to walk me outside. Eric is in a hardcore band called Incarnadine Ashes. Until recently, he was dating Bianca, one of the prettiest and most mysterious visual artists in the school. Eric isn't as pretty as Bianca, but he makes up for his deficiency in the beauty department by embracing his emo side: all dark puppy eyes, passable skills on the guitar. He tells me he's writing a song about me, and we kiss some more. Then I see Jerod's brother.

He's standing twenty feet away and staring at us. He's a freshman at the school, but as a zoned student, a "zonie," I rarely see him around campus—if we happen to cross paths, we don't interact. Maybe we smile. I wait for him to say something and wonder what I should do. He turns and walks away. I know I'm screwed.

Jerod and I have been dating for five months. We're not in love, but we say we are because, at a certain point, it seemed like the right thing to do. We like each other, but mostly we're comfortable. Lately, that translates as bored, at least on my end. We do the same things over and over again: smoke weed, watch *Groove*, have sex, go to parties. We've exhausted any topics we have to talk about. Calling him after school has begun to feel like an obligation.

"Justin told me he's not even hot," says Jerod that night. We're on the phone. I'm lying on my bed, which I've pushed halfway into my closet after removing the folding doors. To my right is a shelf on which I've installed an old desktop computer. It doesn't have the Internet, but sometimes I use it to type up poems and song lyrics. "He's got cystic acne or some shit."

"He doesn't have cystic acne," I say. "Besides, it's not about that."

"What, do you love him?"

"No."

"Then why?"

"I don't know."

"Do you love me?"

"I don't know."

"You don't know?" There is a long pause. "You don't know?"

"I don't know."

We're quiet. I'm aware that this is the last time we'll talk, possibly forever, and I wonder if I've really messed up. I don't want to date Eric Steeler—not really. Jerod has always been nice to me. We've never argued. I like Jerod. So I'm bored—that's no reason to cheat on him. I should have been honest. On the other hand, deep down, I kind of don't care.

"Bye, then," he says, and hangs up the phone.

Lily parks at the far edge of the student parking lot each morning in case there's a need for a midday clambake. We roll up the windows of her black

Volkswagen Jetta and set the air-conditioning to recirculate. She retrieves a glass pipe and a plastic sandwich bag of hydro from the glove compartment and motions for me to select a CD from the case at my feet.

"Pick something good."

It's lunchtime. We've escaped the social pressures of our home table for the privacy of getting stoned in public. Lately, our friends group is antsy: I've been dating Eric Steeler for six weeks and the shine has worn off, but I haven't yet figured this out—I just kiss him less. To add to the tension, after three years of worshipping Lily, our friend Oz finally dated her for a luxurious nine months, only to be dumped for a guy she dated in ninth grade. Now Oz has taken to showing up at her bedroom window in the middle of the night. The more Lily tells him to stay away, the more persistent he becomes.

The first thing people notice about Lily is that she's beautiful. She's a goth porcelain doll: enormous brown eyes, heavily lined; blood-red eye shadow; burgundy hair; petite pinup figure. She wears platform, patent leather, knee-high, steel-toed boots every day. Her resting face is as aloof as a supermodel's, but when you talk to her, you know she's listening. She's loyal. She has a code of conduct that she expects others to abide by: She doesn't share secrets. She expects her privacy to be respected. She's above gossip. She's classy.

I unzip her CD case and flip past Korn, System of a Down, Tool, Slipknot. I make a mental note to come back to Sublime, Cypress Hill, Primus, and Deftones. I'm feeling the shoe-gazey, minor-key drone of the Deftones. I put in *White Pony.*

It's impossible not to sing along with *White Pony.* Chino Moreno uses his voice as an instrument: he's ornamental, but he also screams. His melodic runs wiggle downward, then slide way up to unexpected highs. He racks down to a whisper, then wails.

I listen closely to the vocal distortion in "Elite," and I think about the vocoder Jerod bought just before we broke up. Lily passes me the bowl. She's given me greens.

"I miss Jerod," I say.

"No, you don't," she says. "You just don't like Eric."

I light the bowl. The white-green top layer curls into brown. The chamber fills with milky smoke. I cough and it comes out my nose.

"This is good shit," she says, proud.

When we walk into AP English class thirty minutes later, Mr. Lovejoy knows we're high. Lovejoy is everyone's favorite teacher: he's mellow, but also an intellectual, and he challenges people who he thinks are being lazy, but in a way that's nonjudgmental, even funny. The bonus questions on his quizzes always come from *The Matrix*. He wears a short-sleeved plaid shirt every day, tucked into a pair of khaki pants, and he's never seen without a mug of coffee, resulting in at least one successful parody costume from a student every Halloween.

He holds the door for us and follows us with his eyes as we come into the room. Lily's desk is behind mine, and from the moment we sit down, we can't stop giggling. Lovejoy has seen this before and I know he doesn't like it, but I'm not ready to feel bad about it yet—I feel I have an understanding with Lovejoy. We've just completed a creative writing unit for which I wrote my first short story, "The Fall." It follows two punk teenagers, a boy and a girl, living on the streets of St. Petersburg, who one night decide to sleep on the roof of the State Theatre, where kids go to see punk shows. They score vodka and get drunk, and at the end of the story, the boy falls onto a Dumpster and breaks his spine. I read it aloud to the class, and when I looked up at the end, the expression on Lovejoy's face was unmistakable: he was impressed.

He closes the door when the last student arrives and steps, smirking, to the front of the room. We've been reading *Waiting for Godot* and I'm eager to demonstrate my superior understanding of this text when he says, "How many of you guys smoke weed?"

Lily snorts behind me. Others snicker. A hand goes up, then a few more, and a few more. Lily and I raise our hands. Lovejoy looks from person to person.

"How may of you have done other drugs?"

We all look at one another. Some hands stay up; some go down. I notice some liars.

Lovejoy nods to himself. He looks somber. I turn around to glance at Lily, amused by the line of inquiry. Lovejoy leans back against his desk, contemplating, and before he can ask the next question, Billy says, "What about you, Lovejoy?"

"Me? Do I do drugs?"

"Yeah," Billy says.

Lovejoy smiles. "Drugs are for the weak."

Though he doesn't look at me, I feel he's speaking to me directly. My weed-induced amusement drops down a notch as we begin the day's discussion, and continues to dissipate as our conversation goes on. By the end of the class, I'm sober.

Senior recitals take place at the Palladium Theater, a historical landmark in downtown St. Petersburg. From the wood-and-tile lobby, guests ascend a flight of stairs to the main performance hall and then another set of stairs to the mezzanine. The theater accommodates roughly eight hundred and fifty people. The seats are original leather. Fourteen white marble columns line either side of the room, and a pipe organ fills the wall behind the stage. All of our winter and spring concerts have been here. It's tradition for Gisele, Lily, and me to sneak away from rehearsals and go poking around in the forbidden recesses of the building: our favorite is the narrow room behind the pipe organ, where we can peer at the audience from between the brass pipes. Tonight will be the culmination of every concert, private lesson, jury, theory class, choir rehearsal, All-State competition, and music history lecture of the last four years.

I arrive after Gisele and find her drinking a glass of Riesling in the dressing room, styling her hair. She's wearing a black satin spaghetti-strap evening gown with silver drop earrings, a black choker with an opal charm, and a silver dragon necklace. Lily's gone to a salon and arrives already dressed in a floor-length, spaghetti-strap, white satin gown with her hair styled into an elaborate updo complete with a tiara. I've come in plain clothes and begin to change into my formal attire: a black vintage knee-length cocktail dress with an inverted V-shaped waist, black Mary Janes, elbow gloves, and my mother's gold multistrand necklace. I pull my hair into a bun at the back of my head. I touch up the makeup I'm already wearing: black liquid eyeliner, gray shimmery eye shadow, heavy mascara, and mauve lipstick.

Teachers from the music department are grading tonight's performance. Our grade, worth one whole class, will then be averaged into our overall GPA, and is used to determine our eligibility for graduation from the arts program. I know my teachers don't expect me to sing loudly tonight. Each of them has spoken to me about this privately. I sing beautifully, they say, but nobody can hear me. My performance of "Er, der Herrlichste von Allen" did not impress. We're not singing with microphones tonight, and most of the department, and our friends from school, not to mention our families and our friends outside of school, are coming. I'm at a loss for what to do. I don't know what's wrong with me. It's like something's stuck in my throat whenever I open my mouth to sing. I can't suck in front of all these people.

We're the only three musicians performing tonight. We've alternated the program by language: each of us will sing an Italian aria, then German, then French, then English—then repeat. Our ace in the hole is a song cycle by Gabriel Fauré called *Poème d'un Jour*, during which we plan to pass a red rose from person to person: Gisele will pass it to Lily, and then Lily will pass it to me. My aria is called "Adieu." At the end, I'm to toss the rose into the audience to symbolize the end of the cycle. We'll then finish with our English songs of choice: Gisele's "Jack's Lament"; Lily's "The Black Swan," from Gian Carlo Menotti's 1947 opera *The Medium*; and my "Sally's Song," the last song of the evening.

We wait in the wings. Mrs. Ausubel is at the piano: she's pushing a hundred, but she accompanies like a pro. The curtain pulls back, and we scan the audience without making ourselves visible. The place is packed. Gisele is on first, singing "Il Sogno di Volare," a song from the Cirque du Soleil show *Saltimbanco*, with a lot of high range and feeling: a real crowd-pleaser. Gisele's power, her clarity, her flawless control, all are perfect for this song. With enough luck, Gisele could be the next Anna Netrebko.

She exits stage right and Lily enters. She sings "O del mio dolce ardor" from Christoph Willibald Gluck's 1770 opera *Paride ed Elena*, an aria of love and longing, with lines that slide one into another like brooks into a stream.

Finally, it's my turn. I'm singing "Voi che sapete" from Mozart's *The Marriage of Figaro*, which is fun and lets me show off my high range. I like "Voi che sapete" and I sing it well.

But in my imagined version of this story—the one I still fantasize about, as though doing so will change history—I'm not singing it; I'm singing "Lascia ch'io pianga." I've known "Lascia ch'io pianga" so long I can sing it backward. Before my senior recital, I had disregarded it for years, thinking it represented an earlier stage of my development as a soprano—something I'd evolved past, that was too easy for me now.

In this imagined story, before choosing "Voi che sapete," a song I've learned more recently, I rediscover "Lascia ch'io pianga" in a notebook storing sheet music from my private lessons with Ms. Loup. The notebook still holds my old practice tape—I slip it into my boom box and begin singing along with Ms. Loup's accompaniment. I still know the libretto. Something dormant stirs awake in me that I haven't known was sleeping. It's steady and somber. It builds slowly to a heart-wrenching climax. It speaks of suffering and yearning for freedom. *Lascia ch'io pianga*, it begs. Let me weep over my cruel fate. *Il duolo infranga queste ritorte.* Let my sadness shatter these chains.

Lily exits stage right and I emerge from the left. I cross to the center and look out over the audience. I see hundreds of faces but none of them clearly. I search in the middle, hoping to see my parents, but I can't make them out, so I nod to Mrs. Ausubel and she plays the slow opening chords of "Lascia ch'io pianga." I find the farthest corners of the theater's high ceiling and calculate how much breath I will need to fill them. I drop my jaw, turn off my brain, and let my voice slide out at the right moment. It comes back to me, and I know I've set it free.

After much prodding and many assurances that nothing catastrophic will happen, my parents agree to stay in a hotel for one night so that I can have a graduation party. I invite Jerod without thinking much about it: we've been broken up for five months, so I figure we're cool by now. Ashley comes over early to help me prepare. She wears a red spandex spaghetti-strap tank top and flared jeans; I wear a teal button-down shirt that I've taken in on the sides, fooling around on my sewing machine. We bought food as well as booze because we're trying to be grown-up, and we set up the food on the island counter and pack Smirnoff Ice and Budweiser into coolers on my parents' back porch, where we also deposit several ashtrays. We turn the pool light on so it glows green-blue. We've invited everyone we know. By nine o'clock, the house is full.

I'm on the back porch with Jerod and Sean. Several of my friends from school are here, and Jamie, whom I dated for two months last summer when he was still a virgin. I'm noticing how good Jerod is at mixing with people. I'm also noticing how handsome he is, and how mature he looks with his new light beard and moustache. I take a picture of him as he's talking, and a few minutes later, he picks up the disposable camera and takes a picture of me. We see what it's like to make eye contact while remaining casually far. Later on, we sit next to each other on the floor of my room and listen while Jamie plays Dashboard Confessional songs on my guitar.

Later, we leave out the front door to have a smoke and see three guys I don't know coming up the driveway. The leader is a kid with shaggy hair and a baggy white T-shirt carrying two cases of Natural Ice, one on top of the other.

"Who are you?" I ask.

"I'm friends with Patrick," he says.

"He told you to come?"

"Yeah."

Patrick is a zonie—I barely know him. He has a lazy eye and has a reputation as kind of a thug. Sometimes we say hi when we pass each other in the halls. Once or twice he's sat with us at lunch, but nobody seems to know whose friend he really is. I only invited him to the party because I invited everyone else. I'm surprised he's here.

I look at the cases of beer in this guy's arms.

"What's your name?"

"Kyle."

"Thanks for bringing beer, Kyle."

"All right," he says, and goes inside the house.

Lily and Trent have smoked out Trent's SUV and now are sleeping on each other on the living room couch. Jenna's taken ecstasy she's bought from Patrick; she's outside on the concrete bench by the pool with her head on Jason's thighs as he strokes her hair. Yosi's thrown Rita into the pool, and she's dripping on my parents' upholstered deck furniture. She throws him into the pool in return and now they're both in the pool playing a private game with each other. There's a cluster of drunk sophomores in the sitting room playing strip poker and a group of people in my parents' bedroom testing out the exercise equipment. Kyle is alone in the kitchen, eating broccoli. Someone turns on a Vagrant Records compilation, and the first song is by Alkaline Trio. Everyone on the back porch sings along. We're grabbing beers from the cooler. We're passing a joint.

Jenna's locked herself in the bathroom with Patrick and two of his friends. When she lets us in, the lights are off and they're sitting on the floor with their arms around each other: we snap a picture. Ashley's flirting with my hot neighbor who goes to Catholic school and whom we both want to fuck. My coworker from the Pizza Shack has had too many Smirnoff Ices, and now he's vomiting in the bathroom where Jenna was just rolling with Patrick and his friends—nobody knows where Jenna is now. Gisele's fallen asleep on the couch with Lily and Trent. My coworker's asleep in the guest bedroom with a trash can beside the bed.

A rumor circulates that Jenna is fucking Jason on the grass at the side of the house. Another circulates that the earlier rumor was just a rumor, but nobody wants to believe it. Kyle has projectile vomited down the hallway: the contents of his stomach are four feet long, settling into the carpet and splattered up the walls. Kyle's since disappeared along with Patrick and his friends. There's broccoli in the puke and it's thick like Kyle's been eating crackers, and nobody knows how to get it out of the carpet. Brandi suggests using a vacuum cleaner and I run to the laundry room to get one. On the way, I see Jerod leaving and stop him. He kisses me on the cheek and tells me to call him. Brandi is waiting by the puke when I return. Everyone else has vanished from the hallway. By now, it's after three in the morning—half the party is sleeping or gone, while the other half nurses the last beers on the other side of the house.

We vacuum the puke and the puke gets stuck in the vacuum. The vacuum cleaner is ancient; my parents must have had it since they married in 1984. It's orange and has a particular smell when it's running, like something's burning inside. The more we run it, the more chunks tumble out. Finally, we decide to pick up the puke with paper towels and toss them into plastic shopping bags. Brandi delivers the vacuum cleaner to the porch and the next morning, we sit outside before my parents come home and pick vomit out of the vacuum cleaner with screwdrivers.

It's Friday the Thirteenth, a few weeks after my party. Part six of the Blood on the Turntables DJ series is tonight at the Amphitheatre in Ybor. Jackal & Hyde, Jerod's favorite duo, are headlining. They're hardcore electro: dance music with a hard edge and a dark side. They pull sounds and samples from all corners of the musical landscape and blend them together into grinding beats and complex keys and robotic vocals. I like them more than most of what Jerod listens to—they appeal to my artistic sensibilities. I ride with Jerod in Sean's car, and the sun sets as we cross the Howard Frankland Bridge over Tampa Bay, into Ybor City. Jerod and Sean are discussing freebasing.

"We used to do that shit back in Tampa," Jerod says. "Light it up under the tinfoil." He makes a face.

"That shit is *dirty*," says Sean.

"I've only done it a few times."

"It tastes like ass."

"Yeah."

"Feels good, though," Sean says. He shakes his head. His hair is thinning. I know nothing about Sean, I realize. I don't even know his last name, or where he works now. I don't know that he knows my last name, either.

Jerod laughs and says something I can't quite hear. His friends are meeting us at the venue. They're people I've never met, new since Jerod and I were together. Jerod and I are now friends—friends who flirt and sometimes make out—but we haven't committed to anything serious. We're both seeing other people. We're seeing where it goes.

Sean exits the freeway onto a brick road. He parks on the street near the club and we walk down an alley to Seventh Avenue, which is thick with other club goers. The opening DJ is spinning when we arrive. We push our way to the bar, and then Jerod and I find the dance floor, crowded with bodies. Lasers mounted to the stage slice the air above our heads. Everything is moving. I ask where Sean went and Jerod tells me he's back on the mezzanine with his other friends. So we dance with our drinks in the air, just the two of us. House Wrecka is spinning house, and everyone is getting hammered, jumping up and down, pumping fists, going hard.

I'm alone on the mezzanine looking down on the dance floor. Jackal & Hyde are spinning. I don't know how I got here. I was with Jerod, but I don't see him anywhere. I remember that Sean gave me Xanax when I was down on the dance floor. I remember taking two. But where is Sean now? How much time has passed? I decide to find Jerod. I follow the mezzanine to the stairs and search the faces of people going down, but none of them is him. He's not at the bar. Or on the dance floor. Or in the bathroom.

I find the front door and go out on the sidewalk and light a cigarette. I check my phone. It's been half an hour. I look up and down the street. He's not anywhere and I notice that some part of me doesn't care. I don't care what happens to me tonight. I don't care what happens to Jerod.

I mull this over while I finish my cigarette, then go inside to hear Jackal & Hyde. I decide that I might as well find Jerod, despite how little I may care. I circle the club and find him back on the mezzanine. He's leaning over the banister with Sean and his friends. Jackal & Hyde are still spinning. I touch his back and he turns to face me.

"Where were you?" he asks.

"I don't know," I say.

I get a letter from the college I plan to attend in the fall. It's a school on Long Island that offered me a humanitarian scholarship for the work I've done with my local Food Not Bombs. Now I'm wondering if this was the right decision. They don't have a good photography program, and I'd like to leave photography open as a possible major.

The letter contains the names and phone numbers of my two assigned roommates. It suggests that I call them before we get to school so that we can get to know one another and work out living arrangements. I bring it to my room and sit on my bed. First I dial Lauren in New Jersey. I get an answering machine. I hang up when I hear the beep, unsure of what to say.

I dial Tinsley in Connecticut. Tinsley likes tap dance, she tells me. She loves Broadway music. She wants to major in journalism. She loves Hillary Clinton. I can tell by her high-pitched tone and enunciated consonants that she's a virgin.

"Do you smoke?" I ask her.

"Cigarettes?" she says.

"Sure."

"Sometimes."

"Okay," I say.

MILES MOVES IN WITH HIS BOSS AND HER SON, WHO'S IN A WHEELCHAIR. The house is a mile from Largo Mall down Seminole Boulevard, on a shady street behind the Pinellas Trail overpass. His room is to the left when you come inside. He's furnished it with a queen-sized bed, a dresser, a love seat, and a framed portrait of Derek Jeter, used for snorting cocaine. He keeps a bearded dragon in a tank next to his bongs and a beta fish in a squat glass vase. It's light outside, maybe five o'clock on a Saturday, and we're cross-legged on the rug, Ashley, Clark, Miles, and me. We've each swallowed two beans and snorted half another. Ashley turns on Corina's "Summertime, Summertime," which she prefers to start with whenever she's rolling. We pass a bowl. Miles and Ashley start kissing. The sun sets.

Paul van Dyk's trance *Out There and Back* is on the speakers, graceful and persistent like the paths of orbiting planets. The lights are off, and Ashley is coloring on a velvet black light poster with Magic Markers. I watch her intensely while Clark massages my shoulders. I'm benevolent and joyful. Clark's hands are strong, but he's being gentle. He works patiently, transferring something of himself into me with each gesture: shoulder blades, rib cage, waist, thighs, waist, shoulder blades, shoulders, hair. I lie back against him and close my eyes. Miles turns on a rotating light that throws colors around the room, and I watch them from behind my eyelids:

red, purple, green, blue. The colors want me to love. Clark's hands move over my breasts and a warm light opens up inside me. I'm moving. I'm dissolving. I've entered eternity.

We're all on the bed and Télépopmusik's "Breathe" is playing for the third time: someone's put it on repeat. Everything I touch is impossibly soft. I'm straddling Miles with his back against the wall. Ashley watches us approvingly.

"Can I kiss Sarah?" Miles asks her.

"Do you want to kiss Miles?" Ashley asks me, and I nod. "Go ahead," she says.

Miles's pupils are enormous. I lean into them. Our lips touch and then we're kissing. His tongue finds mine and moves it in small circles. I follow his lead and then I take over, moving his tongue and then sucking on his bottom lip. He closes his mouth. We linger for a moment. I don't remember when I took my clothes off, but I'm touching my breasts while Miles watches at me.

"Can I touch your pussy?" he asks.

I look at Ashley. She heard the question. She's waiting for my response.

"Are you okay with that?" I say.

"Do you want it?"

I hesitate.

"No," she tells Miles.

"Please," he says.

"No."

There's a man in the room and he's in a wheelchair. He's Miles's age. I've never met him, but he looks familiar. He's parked by the love seat like he's been in this room before. There's a song playing that I don't recognize. I climb off the bed to move closer to the speakers, hoping that will help me remember what it is.

I'm cold. A window is open, but I don't see any windows open.

I realize I'm still naked. I don't know where my clothes are. I look around. Miles calls me to the love seat. Next to him, the man in the wheelchair watches me. I lean down to hear what Miles is saying.

"Do you want to kiss Brandon?" he says.

I look at the man in the wheelchair. He's waiting for my answer. Something in him looks lonely and this makes me feel responsible. I lean in and kiss him. His lips are cold.

It's nearing dawn. The four of us are on the bed and we're trying to fall asleep. Clark is holding me from behind. There's something inside me clawing to get out. It's coming through my skull; I can't stand it anymore. I know it's going to kill me. I moan and Clark squeezes me. I hold my face, cover my head. I dig my toes into the sheets and curl my knees up into a ball. I cry softly until I fall asleep.

I wake around noon. Ashley sits on the bed next to me, smoking a cigarette. Clark and Miles are sitting on the love seat. Everyone is silent. I sit up and gather my pants from the floor. I find my shirt shoved down between the bed and the wall and I'm digging around for my keys in my purse when I remember that I rode here with Ashley. I ask her to drive me home. As we're leaving, Clark stops me and tells me to call him later.

In the car, Ashley asks me what Clark and I were doing.

"I was crying," I say.

"We thought you were fucking."

Clark comes over and I lead him past the living room, where my parents are watching television, out to the back porch, where we can be alone. I've brought a lime-green binder filled with sticky photo album pages I bought at Michaels, in which I've stored the photographs I took while I was in New York City. This used to be the binder I kept for music theory class, but I've since thrown away those papers. I've decided that I'm done with music for a while.

We sit on the wicker love seat, smoking cigarettes. I flip through the binder pages, telling Clark the story of each photo. Two days have passed since I learned that we started dating while we were on ecstasy. This is the first time we've seen each other.

"I climbed onto the roof of the hotel to take this one," I say, pointing to an image I took from the ladder of a water tower, looking down at the street.

"You should call it 'Vertigo,'" he says.

I cringe. "'Vertigo'?"

"You know," he explains. "Like when you look down and get dizzy."

I know what the word means—I just think it's a stupid title. I look closely at Clark. He's come in his work shirt. His chin beard needs trimming. His complexion is shiny. He's stocky and muscular—not ugly, per se, but there's also nothing attractive about him.

"Maybe," I say.

He takes the pen I've hung from the collar of my shirt, lifts the plastic from the photo album page, and writes "Vertigo" beneath the photo, then lowers the plastic.

I walk him to the door when it's time for him to leave. He leans down to kiss me and I step backward, indicating that I don't want to kiss him while my parents are home, that they might see. He nods in understanding and kisses my forehead instead. That night, I call him to say we're breaking up.

I'm leaving for college in a month. I'm in my father's car with Gisele, and we're driving around Largo, looking for someplace to smoke this joint. It's dusk. We pass the Denny's on Missouri Avenue where we fed our hangovers all through high school. We pull up to a red light on the corner of Missouri and East Bay Drive. To our right is McGill Plumbing with the giant faucet affixed to its roof from which neon drops of water fall, one after another. We turn left down East Bay and pass the glass brick clock tower at the corner of Largo Central Park, and continue on toward the adjacent Largo Cultural Center, where we turn right toward the library and pull another right into the empty parking lot. We climb out of the car and walk toward the park's public restrooms, which we aren't aware closed at nine.

We're surprised to find that the doors are locked. This throws a wrench into our plans. We walk around the perimeter of the restroom structure assessing our options. There are ventilation slats near the ground but neither of us can fit through them—they're too narrow. There are more near the roof but they're too high for us to climb, and neither of us trusts our ability to fit through those, either: they're shorter, and square. Both Gisele and I are small, but Gisele is smaller and more nimble. We decide the riskiest route is our only bet. I make a cradle with my hands and boost Gisele up to the ventilation window above the door, which is about a foot high and two feet across. She shimmies through it, drops easily to the floor, and unlocks the door to let me inside. I lock it behind me. We high-five. We're proud of ourselves.

Lately I've been turning my old pants into purses. I cut the legs off and sew up the bottoms, then make straps from old belts and bedazzle the fabric with patches and studs. The one I'm carrying tonight is navy corduroy with a teal canvas belt and red lace flowers on one side. It's cool because the old pockets become hiding places for things like gram bags and lighters. Right now, I'm carrying two grams of weed as well as the joint I'm currently lighting. Gisele's walking around the restroom testing the acoustics of different corners. She's in the process of preparing an audition tape for Cirque du Soleil. She's singing "Il Sogno di Volare," which she plans to put on the tape, at the top of her lungs. She sounds terrific.

We're finishing the joint when we hear jingling. We've been talking and singing for the duration of our smoke-out, songs we sang together four years before in the Renaissance Festival, and songs we learned in choir, for which we harmonize: I take first soprano and she takes second. We shut our mouths and the jingling passes. We let a few moments elapse while we stare at each other, and then we start to talk again, tentatively, keeping our voices low.

We're back to singing when the jingling returns. This time it walks right up to the door and knocks on it, three times, hard and demanding like it wants something from us. "This is the police," it says. "Open the door."

Gisele mouths "fuck" and we scramble from the floor. My heart feels like a trebuchet launching boulders against my breastbone. I nearly pass out as I reach for the door handle, my periphery closing in, my face numb, my palms sweating. I'm fucked. I'm so high. They'll know how high I am.

Outside, two cruisers are parked in a V-formation barricading us against the restrooms. Four officers stand with their hands on their hips. One holds the leash of a German shepherd. The German shepherd pants, staring at me.

"Let's see some licenses, ladies," says the officer on my left. They're all young, and white. We reach into our purses and retrieve our wallets, struggling to contain our shaking. I'm paranoid that moving my purse will release the smell of the weed inside it.

"These restrooms are closed," says the officer, examining my license. He passes Gisele's nonchalantly to his left. "What were you ladies doing in there?"

"We were singing," Gisele says.

"We're singers," I explain. I say it before I realize I'm saying it.

Something about the way they're looking at me makes me want to tell them everything. I feel my purse inside my hand. I avoid looking at the dog.

"How'd you get inside?" he says. He passes my license to the officer on his left, who takes both of our licenses to his cruiser. I watch him walk away.

"Through the window," says Gisele.

"You climbed through the window?"

We nod. He looks at us.

The other officer returns. He shows my license to the first one, who looks at it again, more closely. They look at each other.

Here's where I should explain that my mother is at this time a city commissioner. She was elected first when I was fifteen, and to a second term earlier this year. She became involved in local politics while working for the Largo Police Department as a victim's advocate, a position for which she was hired shortly before I was born.

The first officer hands our licenses back to us. We replace them in our wallets and await his verdict.

"The park restroom closes at nine," he says.

By the end of the summer, Jerod and I are officially back together. I've decided that I love him but that this relationship won't last after I move away. Besides, he's been doing too many drugs. Last weekend, he went through thirty-nine nitrous poppers in a single night with two of his friends: I told him I was surprised he was coherent the next morning, but I didn't tell him not to do it again. I'm afraid that if I said this he'd call me a hypocrite. He'd be right. I do drugs, too. But there's something different about the way he does them.

Gisele and TJ have been dating since last summer. Tonight he's throwing a party at his apartment at which he'll be spinning, and he's also invited Jerod to spin. When we show up it's still light outside and we find him spinning with no one else in the apartment except for Gisele, who's in the kitchen fixing herself a drink. TJ shares an apartment in unincorporated Pinellas County with a guy named Geronimo who's rumored to be some kind of guru. Geronimo is our age, and Peruvian, with hair that curls down to his shoulders and a way of leaning in while you talk that makes you feel like you're saying something deep and soulful. Their unit is on the second floor of a building in a new complex, and has a balcony that overlooks a pond inhabited by alligators. I sit on the new couch, waiting for the party to start, and it strikes me as I sit there soberly that this, this silly party, is the thing I've arranged to do with my evening. I wonder what else I could be doing.

Jerod and TJ talk over the turntables. They're comparing the strengths of electro versus breakbeats when TJ pulls a DJ Baby Anne record from a crate at his feet and sets it spinning on one of the turntables, turned all the way up so that I can't hear what they're saying anymore. I watch their lips moving, and I imagine that the sound between us has become a wall, a wall they've erected to keep me out.

Geronimo comes home and stands in the hallway entrance. He's bobbing his head with the music, and as I watch him, the music comes into focus and I hear Baby Anne clearly. The Bass Queen. Electro, funk beats, and Miami bass: this is what Florida breakbeats is all about. Generally I like Baby Anne, but not right now. Right now, the sound is wrong. The pressure of the bass makes me feel as if I'm sinking, and the deeper I sink, the lonelier I feel. I don't want to roll tonight, I think. I want to go home.

The song ends and Geronimo steps into the room. "Who else is coming?" he says. "Let's eat these beans." So we do.

My oldest friend, Kelsey, invites me to a party near my parents' house in Largo. I've known Kelsey since we were two. We grew up a few streets away from each other. As a kid, I spent almost as much time at her house as I did at my own. Kelsey's family is my family; her brothers are my brothers. She's the closest I've come to having a sibling.

The people throwing the party are her friend Logan and his roommate, a guy she's never met, but Logan is cool, she says. It's a small group. Real chill. She picks me up in her gold Jeep Cherokee. I wear a knee-length, pleated skirt paired with a black argyle sweater vest, gray flip-flops, and heavy eye makeup. I'm looking like a de-virginized schoolgirl. The night is hot and sticky. We drive to a dead-end street off of Walsingham Road, near Heritage Village, where we ate raw sugarcane on field trips in elementary school. There are no curbs on this street; the concrete mixes with grass in a jagged line ending at three red diamond-shaped signs, where cars are parked in the grass near a beige duplex. This is our destination.

A few hours later I'm at a round table, surrounded by strangers playing cards in an otherwise stark living room: white tile floor, no sofa, an empty kitchen on the other side of the room. I'm sloppy drunk. We've been playing this drinking game since we arrived. I'm sitting next to Mitch, whom I've just met. Mitch, with his tight black T-shirt, tattoos, and baby face, who's teasing me about my outfit, whose bedroom is twenty feet behind us. He's a line cook at a fancy restaurant on Indian Shores. He hates his father. He's a year older than me but dropped out of school in eighth grade. He has a way of looking at me sideways that makes me want to impress him, and I can tell he knows this—he wants me to put in the effort. We're talking about our love of hardcore and post-hardcore. I'm exaggerating my own feelings on the subject. Poison the Well is amazing, I say. I love Atreyu, and mewithoutYou.

"I need someone to go to the Hopesfall show with me on Friday," Mitch says.

"I'll go," I say.

Hopesfall is playing at the Masquerade in Ybor with Every Time I Die, the Beautiful Mistake, and Celebrity. I've listened to Hopesfall but not very closely—just enough to put them on a mix CD I made for Jerod last summer when we started dating the first time. They're a lot like their contemporaries: a blend of emo and hardcore punk with ample screaming. I wear my red University of Hawaii T-shirt with ripped jeans that show off my ass and red Converse, the toes of which feature tiny skyscapes painted by the girl I started dating in rehab, for cutting, the year before. I arrive alone. The Beautiful Mistake is playing. Mitch is standing at the back of the theater near the bar, on a raised area surrounded by a banister. This is our first time seeing each other since his party. He smirks at me.

"I didn't think you'd come," he says.

I smile and turn my attention to the stage. I feel mysterious and sexy. I've gotten his attention.

"Have you heard these guys?" he says, referring to the Beautiful Mistake.

"No," I say.

"They're great."

"Cool."

I listen to the band. They're nothing special. But they're also not terrible. I bob my head and look interested.

I'm meeting Mitch at the Steak 'n Shake with his friends, whom I discover know many of my friends. I'm riding around in Mitch's truck, listening to hardcore while he pounds his fist on the steering wheel and refuses to change the CD to something less angry. I'm meeting Mitch's aunt in her salty wooden condo on the beach. I'm taking color-tinted photographs of Mitch from above with my Lomography camera, straddling him on his bed. I'm buying him used copies of Kurt Vonnegut books, encouraging him to read them. I'm watching *Harold and Maude* on his roommate's bed while Mitch spoons me, teasing him for loving only hardcore with the exception of Cat Stevens. I'm sleeping with Mitch for the first time.

I meet Jerod's mother at his GED graduation ceremony in mid-August. She drives in from Tampa and we convene at dusk in the parking lot of the Bayfront Center arena in downtown St. Petersburg. Jerod's father and brother have gone inside. I'm wearing the same outfit I wore to Mitch's party: knee-length, pleated gray skirt and black argyle sweater. I feel respectable. I don't know what Jerod has told his mother about me; I don't even know if she's aware he has a girlfriend. She's petite and soft-curved with strawberry blonde hair. She hugs me warmly. She smells like peaches.

Jerod's best friend from Tampa is here, Desi. I've heard about Desi. I once listened to Jerod give her directions over the phone for what to do if she snorted heroin and wanted to throw up (just throw up). She and Jerod have never dated, but they've slept together at least once. She's pretty. Really pretty. Her blonde hair is pulled back so tightly it's almost sheer. She's drawn her eyebrows in high and dark, and lined her lips with brown liner, then filled them in with light pink lipstick. She's wearing cutoff shorts and a skin-tight Bebe T-shirt and running shoes. Jerod's mother didn't tell him Desi was coming. She's a surprise.

His mother's also brought another of Jerod's friends, who's brought her baby, whom Desi is holding. I know little about Desi, just bits Jerod has told me, but as I watch her making sweet faces at this baby, I think of the one thing I know that stands out: Last year, she was at a party thrown by a close friend. While she was drinking in his kitchen, he grabbed her head and slammed it into the counter, and she passed out. When she awoke, he was raping her.

"What was that?" she sings to the baby, raising her eyebrows. "What happened?"

That night, Jerod calls me to say he needs a break. "You're leaving," he says. "It's too much for me." He's crying.

"You need some time?" I say.

"I do."

"How long?"

"Two weeks," he says.

I calculate forward. "I'm leaving in two weeks."

Ten days go by; I'm leaving for college in three days. I'm lying on Mitch's bed in his dim bedroom. The closet doors are open, revealing duplicate black and white T-shirts, the only colors he wears. Band posters hang on the wall. An electric guitar leans against a stand. It's after midnight. I've told my dad I'll be home by one o'clock, so I ask Mitch to take me. Instead, we continue talking, which leads to kissing, which leads to his hand in my pants. He begins to take my pants off.

"We don't really have time for sex," I say. "I have to go home."

"We can do it fast," he says. He continues taking my pants off. I say nothing. We start making out. Soon, he's on top of me. He has his hands on my shoulders.

"I need to go home," I say.

He kisses me. I kiss him back. Then he's inside me. He's thrusting.

We're having sex for a minute when I say it again. "I need to go home," I say, pushing his shoulders. "Stop it. Stop."

He doesn't stop. I look up at him. He's looking directly at me. There's no way he doesn't hear.

"Stop," I say. "I need to go home."

He continues to ignore me. His expression is intense, unflinching. I push his shoulders away and lift my knees, put my feet on his hip bones, and kick hard. He falls backward. I scramble upright.

We're silent on the way to my house. Mitch stops at a red light in front of the pizza parlor where I worked until recently. I look over at him and he looks back at me. I feel like I should say something, but I don't know what. I don't know what just happened.

"I would never have done that if I'd known," he says.

That night, I start a new diary in the book Ashley gave me when she learned I was moving to New York. On the cover is an illustrated picture of a lady in a short red evening dress leaning on the hood of a taxi with the Statue of Liberty in the background. The red dress is filled in with shiny red beads. The image is reproduced in the lower left corner of each pink interior page. "I'm not sure if you would consider it rape," I write. "By definition, I guess it was, but it wasn't exactly how I imagined rape." I write three more pages, telling the story of what happened, and then I realize I left my cell phone at Mitch's house. "Shit. I'm going to call him," I write.

There's an ellipsis; then I return: "He said he was sorry and that he 'didn't mean to force anything upon' me. I know I'm going to sound incredibly weak saying this but his behavior after what happened portrays an extreme sense of remorse. I really think that he didn't realize what he was doing when that happened."

The next morning, I drive to Mitch's house on the way to Jerod's. The front door is unlocked. Mitch pokes his head out of the bathroom when he hears it open. He's holding a comb in his left hand. We stare at each other. I walk silently past him. I find my cell phone sitting on the dresser in his bedroom. I leave without saying good-bye.

It's the night before I leave Florida for good. I'm stoned and sitting on the driveway with five of my friends, including Ashley and Gisele, and we're talking about sex and where we'd like to have it. Yesterday, I went to Jerod's house. I cried, and so did he, and then we fucked, and then I left, and that was that. Now I'm saying I'd like to fuck on a pile of clean laundry in a Laundromat at midnight with someone I just met. Ashley wants to do it onstage at a strip club. Gisele wants to do it in the woods on a bed of roses.

Mitch was supposed to call me tonight on his way back from the Shadows Fall concert, but instead, he explains later, he fell asleep. He had wanted to talk before I leave. I'm not mad about it. I have a picture of him in my diary that I plan to tape to my wall when I get to college. That night, I write: "I was thinking the other day about how humans, to my knowledge, are the only living things that intentionally mark time. We create tactile memories for ourselves with photographs, music, memorabilia, and in my case, cutting. Perhaps that's another reason why we're a superior species. Or perhaps, that's why we're not."

Epilogue

I last see Jerod five years later, soon after graduating college. He messages me on Myspace. It's pure coincidence that I even see the message: I've long since stopped using Myspace and have signed in to hunt down an old picture. He's included his phone number in the message, so I invite him to see a movie with my friends and me that night.

I drive my gold '93 Cadillac DeVille to his dad's house, where he's living in an apartment above the garage, to pick him up. He's been out of prison for a few weeks following his third arrest, for trafficking cocaine, and he shows me around his small, dim space. He has a beige sofa and beige carpet, a wooden coffee table, a two-person Ikea kitchen table, and a set of turntables—he's still spinning. It's impeccably clean. We sit at his kitchen table drinking water, and he tells me about his other two arrests: the second for possession of Oxycodone, and the first for domestic violence. He'd been living with a girlfriend, he says. She cheated on him, and his friends had to stop him from hitting her. Even though they stopped him, the police arrested him anyway stating that he would have hurt her if they hadn't.

Standing outside the theater later that night, I learn that he's dating a new girl. She's currently in prison. They were arrested together for trafficking cocaine, and she took a fall for him because he's still on probation for his last arrest. He hadn't asked her to do that. His guilt is overwhelming. He gazes across the streetlamp-lit parking lot, shaking his head. My friends have gone inside, but we don't care. He doesn't know what to do, and I don't know what to say. Knowing Jerod, I'm sure this isn't the end of the story. I'm sure that, like me, he's going to take it as far as he can, but I can't tell him that. So I say it will be okay.

After spending five years with Miles, from ages seventeen to twenty-one, Ashley enlists in the army to get sober. She hates every minute of basic training. She writes me letters in which she describes the terrible food and the grueling schedule. She misses her mom and her freedom. She's made a mistake.

Two months after enlisting, she cracks her pelvis during exercises and returns home to recuperate. She takes up partying again right away, drinking and doing pills while her mom spends her final weeks in bed. One night, she climbs onstage at a strip club and dances naked with a tampon string dangling from her vagina. Weeks later, just days before her twenty-second birthday, Ashley calls to tell me her mom has died.

Later that month, when I'm home for a long weekend, I pick her up in my mom's red Corvette and we drive to the state fairgrounds, eating candy straws on the way. As soon as we arrive, we buy margaritas. We go on the Zipper, the Hurricane, and the Gravitron, twice. We puke into trash cans, laughing at the messes of ourselves.

Ten years later, Ashley lives in Largo. She's training to be a nurse. Her daughter is four. She's clean.

Gisele spends the years after high school learning to dance with fire. She apprentices with fire dancers she meets down on the beaches. They teach her poi and flaming rope, hoop and fan. She moves to Orlando, where she attempts to make a name for herself. She marries; then she divorces. She moves back to Largo, where she joins up with an acrobalance troupe, with which she still performs. She works for an electrical engineering company. She plans to return to school to study physics.

Eight years after leaving Florida, I pass Mitch on the street. I'm living in New York. I've just finished a master's degree and am working at a bookstore and writing a novel. I'd heard a rumor some time before that Mitch was moving to the city, but I haven't looked into it much—I try not to think about him much these days. Now he's standing ten feet away from me, outside of a tattoo parlor in lower Manhattan, smoking a cigarette. He waves me over. He's apprenticing here, he says, but he plans to move to a new parlor as soon as possible. His arms are now covered in tattoos. He's muscular beneath his shirt.

We'd kept in touch a bit while I was away, and after my new boyfriend cried when I told him what had happened between us, I'd come to the conclusion that it was insignificant and that I should not make a big deal out of it. So I'd told no one else. Not even my husband.

The last time I saw Mitch was in the summer before my sophomore year of college. He'd come to pick me up at my parents' house one afternoon when the sun was blistering. I had taken to wearing a different knit beanie every day, regardless of the outfit—something of a security blanket. My beanie that day was sky blue and furry, the yarn shimmery. I'd paired it with a pink plaid dress over ripped, drawn-upon jeans and backless black-and-white Converse All Stars.

We stood on my parents' new hardwood floors with the front door open, his truck idling in the circular driveway. He appraised my outfit.

"Why do you insist on making yourself look like shit?" he asked.

I stood there stunned. He faced me, waiting.

"Actually, I don't," I finally said. I told him to leave. Even now, I think of his face as he turns toward the door. The effect of his words washes over him and becomes real. His expression changes from amusement to remorse. The words "nice guy" spring to mind. I resist this.

Talking to him outside this tattoo parlor in New York, I remember the watercolor sets in his bedroom and his designs for future tattoos. They were graphic and haunting, drawing upon his metal aesthetic, his propensity for aggression.

"We should keep in touch," he says. "Maybe hang."

I nod. We finish our cigarettes. I leave.

A YEAR PASSES, and I decide I want to cover a friendship tattoo I got when I was twenty-three, with a friend I don't speak to anymore. We met in fifth grade, when I transferred into her school to enter the arts program. She was freckled and blue-eyed, like me, but more outgoing. We hated each other at first—we were always competitive. Then Mrs. Dorff sat her next to me during an art lesson and she asked to borrow a glue stick. After that, we became inseparable. We remained best friends for seventeen years before our latent animosity raised its head.

The tattoo is two-parted: a drawing on my right hip of two stick figure girls holding sticks of dynamite, identical to the one on her left hip. Beneath hers, the word "Forever"—beneath mine, the words "& ever." I no longer want to face daily this consequence of our reuniting after one of many periods of not speaking to each other. It's too painful. I ask Mitch to cover it up, thinking that he'll give me a good deal. I'm also curious to see what the dynamic between us will be. I have yet to define for myself the nature of our last sexual encounter.

I'm meeting him at his new tattoo shop. I'm seeing, for the first time, the tattoo he's designed for me, and I don't like it. I asked for an elephant, but I wanted something subtler than what he's drawn: I want black and white with shading. Minimal lines. High detail. Something soulful, gentle. I want to cover the dead thing up with something living. I think to myself that I should have remembered his style: hard lines, bright colors, high contrast. He's drawn me a cartoon. I don't know what to do now. I blame myself for failing to foresee this.

I'm saying yes anyway, not wanting to insult him. I'm lying prone on the chair. I'm feeling the cold of the alcohol cloth he uses to sterilize the area. I'm smelling the burn of the fumes. My pants are unbuttoned and pulled down below my hips along with my underwear. My shirt is above my waist. I'm still. My husband sits on a chair next to me but a few feet away, out of reach, out of sight unless I crane my neck.

Instead I watch Mitch. I feel Mitch's breath on my skin as he leans close to me. He touches me with warm black-gloved hands and pulls my skin taut. When he brings the tattoo gun down on my hip, it stings and I wince. He looks up.

"This is a tough spot," he says.

I want him to show me he cares about my pain by looking sorry. To shake his head and apologize, rub my skin to help the pain disperse outward. I want him to ask me if I'm okay. I want him to touch me again with the gun.

We take a break midway through to smoke a cigarette out back. My husband comes with us. I won't tell him about our history for months, and then he'll ask why I got the tattoo. I won't have an answer.

We go back inside and finish the tattoo. Mitch asks me what song I want to hear, and I ask him to put on Social Distortion's "Ball and Chain." I hum along with Mike Ness's folk-punk melody while thinking about his lyrics: A plea to take away a lifetime of suffering. An admission that wherever he runs, he finds himself. Mitch fills in the shadows between the foliage around the elephant with thick black and green geometric blocks. He tints the elephant's trunk with a baby pink tip. The eye is a blank white circle with no expression. The elephant leaps and smiles. Mitch stops to wipe the blood periodically when it trickles down my side. He lifts the gun when I'm hurting too much.

As the elephant heals, it forms a thick scab that cracks when I stretch it. My hip swells and is painful to touch and lie on. I'm gentle in the shower and clean it twice a day with a mild, sterile soap. I massage the elephant with moisturizing ointments and cover it with gauze. Eventually, it smooths over, becomes integrated.

The Mayor of Williams Park

At the same time came the disciples unto Jesus, saying,
Who is the greatest in the kingdom of heaven?
And Jesus called a little child unto him,
and set him in the midst of them,
And said, Verily I say unto you, Except ye be
converted, and become as little children, ye
shall not enter into the kingdom of heaven.
Whosoever therefore shall humble himself as this little
child, the same is greatest in the kingdom of heaven.

Matthew 18:1–4

A line forms outside the Trinity Lutheran Church in St. Petersburg, Florida, each Saturday morning at six o'clock. Breakfast is at nine, and by that time the line will have snaked halfway around the brick building, down the narrow alleyway lined with Dumpsters and yellow-tinted, low-income apartments, and back around to where it began, on Fifth Street. For those three hours before the doors open, the people in line smoke and talk, sharing stories of where they slept the night before and who or what awoke them and forced them to move. Which of their friends and family is missing or dead, who is housed, if anyone, even temporarily. Who was arrested. They're clean or dirty, black, white, Latino, sane or impaired, pregnant, disabled, old, young—and poor.

At six thirty, a two-hundred-fifty-pound, six-foot-tall black man arrives and cuts to the front of the line. He breathes heavily and walks with difficulty, looking down at the ground or straight ahead. He wears a fedora and a clean button-down shirt. He is G.W. Rolle, the man who's always on their side. They move to let him pass. Some greet or hug him. Some ask him for money or a smoke. He lingers while he finishes his cigarette, opens the church, and goes inside. He takes the elevator to the fellowship hall on the third floor. He goes to the kitchen and starts cooking.

I arrive at eight o'clock on a warm July morning in 2015. G.W. is in the kitchen with four volunteers, all homeless or formerly so. Each cooks a part of the morning's breakfast, frying eggs or sausage, seasoning home fries, cutting fruit salad or squares of cream cheese. On the way in, I pass a tall rolling rack of donated pastries, and two long plastic tables under a row of windows, and two boxes of plastic gloves. I've let G.W. know I'm in town to write a story about him, and he's told me to come to the breakfast—he'll have me serving grits. G.W. believes in service. When he got off the streets in 2006, he turned back around to help others up after him. Not everyone does that.

G.W. is an ordained minister whose church, Missio Dei, has an office in the basement of Trinity Lutheran. Missio Dei describes itself as "an imperfect church of imperfect people inviting other imperfect people to find the perfect love." When asked his denomination, G.W. says, "I identify as the hands and feet of Christ." Each Saturday morning, he serves a free breakfast in the fellowship hall; each Sunday night, he serves a free dinner and gives a Missio Dei service in the small wooden sanctuary. Attending services isn't mandatory, but in the pews there always sits a scattering of people from the streets. Afterward, they eat.

On this warm July morning, when the breakfast is ready, we circle in the fellowship hall for a prayer. Outside, people wait at the brick edge of the parking lot. On the tables behind me sit large plastic serving trays of eggs, grits, home fries, sausage, fruit salad, four different baked goods, cream cheese, and squeeze bottles of jelly and syrup. On the rolling rack, loaves of bread have been wrapped for people to take with them when they leave.

"We got a lot of people here," G.W. begins in his baritone. "As I always say, it ain't hard to start, it's hard to finish. If people wipe the tables, sweep the floor, you know, carry the coffee back and whatever, we would be out of here in a shake of a lamb's tail. That's it," he says. "I've been criticized for talking too long." Everyone laughs.

Pastor Tom Snapp of the Lutheran church takes over. "We thank you, heavenly Father, for bringing us here this morning. We thank you for being able to cook and serve. We thank you for the money you provided for us to buy the groceries. We thank you, Lord, for everything that's so gracious that you give us. Help us to be humble. Help us to serve those who are hungry, that they may, through the food we serve, find the joy and grace of your love. In this we pray, in Jesus's name. Amen."

"Amen," we echo.

In 2015, the state of Florida was home to an estimated 35,900 homeless people according to the annual point-in-time count conducted by the federal Department of Housing and Urban Development (HUD). On a single night in the last week of January, volunteers in each county across the country count the numbers of sheltered and unsheltered homeless people living there. Counties report their findings to

HUD, broken out into categories of race, age, gender, familial status, and disability. In 2015, Florida had the third-highest number of homeless people in the country, behind New York and California. It had the second-highest number of unsheltered homeless. It had the highest number of homeless veterans. That year, Pinellas County, which sits on a peninsula near Tampa on the west coast of Florida, was home to 6,853 homeless people; 40 percent of them were children.

G.W. is a member of the Pinellas County Homeless Leadership Board, which is responsible for distributing Continuum of Care money from HUD each year into programs for permanent or transitional housing, supportive services for unhoused individuals, shelter data infrastructure, and homelessness prevention programs. Other members of the board include local elected officials, county sheriff Bob Gualtieri, members of the Juvenile Welfare Board, public service experts, members of local faith-based service providers, and one homeless or formerly homeless individual. Currently this is G.W.

I first saw G.W. in a 2006 documentary called *Easy Street*, released by St. Petersburg–based production duo Wideyed Films. The film documents one year in the lives of five people experiencing homelessness in St. Petersburg, the largest city in Pinellas County, which had a reputation for being "a great place to be homeless"—with mild winters and abundant help agencies and free meal programs. The subjects include Patrick, an alcoholic; Peg, a victim of domestic violence; Karl, a bipolar teenager; Jaime, a widow who's evicted along with her two children; and G.W.

"My name is G.W. I'm forty-nine," comes his voice over a black screen. "I've been on the streets on and off since I was fourteen."

The scene opens onto a busy four-way intersection on a sunny day. G.W. stands on the median in a fluorescent yellow *St. Peters-*

burg Times T-shirt. He holds aloft a copy of the newspaper from the stack at his feet.

"I'm still selling papers. I'm doing it every second," he says. "But it's not gonna get you anywhere."

We cut to a shot of G.W. in Williams Park. He wears a long-sleeved black shirt with a skeleton screen print. He sits on a bench with his arm slung over the back.

"A human being can get used to anything," he says. "My friend asked me—he's a religious guy—and he says, 'What does Judgment Day mean to you?' And I thought for a minute and I said, 'To me, Judgment Day would be me standing before God, and God telling me that, "I gave you a certain amount of intellect. I gave you writing ability, communication ability; I gave you cooking ability; I gave you the ability to get along with people; I gave you this very deep voice—and you haven't used any of it."' To me, hell is squandering my abilities."

He was seventeen and living in New York City when he and his friends robbed a drug dealer and the dealer's girlfriend was shot. She reached for G.W.'s gun, which neither knew was loaded. G.W. was convicted of manslaughter and subsequently spent five years in prison.

"A conviction of manslaughter stays with you for the rest of your life," he says, looking out over the park. "The good thing was I got a scholarship to Syracuse University."

At Syracuse, G.W. majored in philosophy and English, with a minor in religion. He wanted to be a lawyer or a teacher. "But it just couldn't be," he says. Everywhere he went—California, Washington—once they saw his record, they never called him back. "The one mistake has been that five minutes when I was seventeen years old," he says. "My life has not been the same. Well, it has been the same—it's been shitty."

G.W. has been arrested more times than he can count. He's been in LA County jail. Orleans Parish prison. Rikers Island, when he was a kid. Every county jail in Buffalo. Most of the time, it was simply for being homeless. "I wrote about it in my book," he said.

His autobiographical novel is called "Things That Break." It tells the story of a formerly homeless minister, also named G.W., who presides over the citizens of Williams Park, and his entanglement with a particular homeless family called the Coyles. It opens with a prologue that details a fictionalized account of G.W.'s last arrest, for urinating in the middle of the night behind a Dumpster near where he'd been sleeping. He had a repeat work ticket for a day-labor gig the next day, which he'd hoped to turn into something steady. "At the end of the week," he writes, "I'd be able to get a room for seventy-five dollars. Then bye, bye streets. If I got the steady." Because of his arrest, he couldn't get to work the next day. "Bye, bye job, I thought. Just like that."

After "Things That Break," he plans to write two essay collections and finish a mystery novel he began the first time he went to prison, over forty years ago. "I'm on the bottom, but I'm an intellectual," he says. "Most of the people on the bottom aren't intellectuals."

G.W. has always been a writer. He began writing during his first incarceration: short stories, essays, and a few attempted novels. When I met G.W. in 2009, he was putting together the second issue of *The Homeless Image*—a street paper by and for local homeless, to be sold by the homeless population to the general public. Then, in 2010, the City of St. Petersburg passed an ordinance banning the roadside sale of newspapers, and *The Homeless Image* folded.

"I'm a better writer than Richard Wright and James Baldwin," he says, adding that James Baldwin had the "advantage" of being black and gay. "Baldwin had angst." He laughs. "He didn't mince words."

In November 2015, G.W. was scheduled to present a paper in Geneva, Switzerland. The report, called *Cruel, Inhuman, and Degrading*, was published by the National Law Center on Homelessness and Poverty, where G.W. sits on the board of directors. It concerns the criminalization of homelessness in the United States. He wasn't there to present it. Instead, G.W. spent a night in Los Angeles County jail after turning himself in for a twenty-seven-year-old warrant for drug possession. The State Department had denied him a passport after the warrant came up in his background check—which is how he first learned of it. When he went down to take care of the warrant, he expected the statute of limitations to have long-since expired. But when he stepped outside to smoke a cigarette, he was approached by three officers. They booked him.

Cruel, Inhuman, and Degrading was published in 2013 based on research carried out by Yale Law School and Law Center staff. It outlines numerous violations of the International Covenant on Civil and Political Rights and the Convention Against Torture carried out against persons experiencing homelessness by the US government at both state and local levels. It cites, among other things, lack of shelter space and available services for families in crisis; lack of affordable housing; the disproportionate representation of nonwhite and disabled people within the US housing-unstable population; and local ordinances, such as those against panhandling and sleeping outside, which target unhoused people. These violations specifically affect homeless people, criminalizing them for being homeless or for performing activities associated with homelessness. A Law Center study published in 2014, called *No Safe Place*, showed that these sorts of criminalizing ordinances had been sharply on the rise since 2009.

Though this topic has been circulating through the national conversation since 1999, it has gained traction in recent years. Many cities nationwide have lifted ordinances against panhandling, either voluntarily or by federal court order, citing the First Amendment right to free speech. In 2015, the Department of Justice filed a brief discouraging law enforcement from punishing people for sleeping outside if there is nowhere else to go, citing the Eighth Amendment against cruel and unusual punishment. They declared that "needlessly pushing homeless individuals into the criminal justice system does nothing to break the cycle of poverty or prevent homelessness in the future," and doing so "further burdens . . . scarce judicial and correctional resources." Soon after, President Obama incentivized homelessness-prevention institutions and outreach programs by making it more difficult for municipalities to obtain HUD money if such programs aren't in place, or if they aren't moving to reduce laws criminalizing homelessness.

Instead of lifting ordinances targeting homeless individuals, some localities have put money into jail-diversion programs or shelters such as Pinellas County's Safe Harbor, which is managed by the Pinellas County Sheriff's Office and stands on the campus of the Pinellas County jail in a former low-security annex.

Safe Harbor opened in 2011 in response to a rise in the homeless population and overcrowding in the Pinellas County jail, where many of them were ending up. Clashes between the homeless community and local law enforcement had reached a fever pitch in recent years, especially in St. Petersburg: In 2007, just days after two homeless residents were found murdered, St. Petersburg law enforcement raided two camps, or "tent cities," seizing property and slashing the residents' living spaces with box cutters.

That year, Pinellas-Pasco public defender Bob Dillinger announced he would no longer represent homeless people arrested for violating municipal ordinances in St. Petersburg to protest what he called the "excessive arrests of homeless individuals." Of the 879 people booked into the Pinellas County Jail on such ordinances over the last year, 676 came from St. Petersburg, and the vast majority of them were homeless. Without the public defender to represent such cases, the city would be forced to hire private attorneys or face a class-action lawsuit. Private attorneys would charge the city considerably more than the public defender's flat fee of $50. "They're trying to get the homeless out of the downtown area," Dillinger told the *St. Petersburg Times*. "I would rather have them spend money on helping the homeless rather than arresting them."

By 2009, the city had passed six new ordinances criminalizing homelessness, and the National Law Center on Homelessness and Poverty, along with the National Coalition for the Homeless, named St. Petersburg the second meanest city in the country for homeless people. Soon after, advocates filed a class action complaint against the city on behalf of local homeless individuals "who were routinely penalized for using public space to perform basic bodily functions when they had nowhere else to go," among other measures.

Desperate for a solution, in 2010 the City of St. Petersburg hired Robert Marbut to advise them. Marbut is a self-proclaimed, self-taught "homeless expert"—a Texas community college professor turned independent consultant peddling his Seven Guiding Principles for Transformation method of solving homelessness. Marbut's principles, which he calls his "Velvet Hammer," cast community residents as codependent enablers to loved ones addicted to homelessness, with phrases like "Free food handouts and cash from panhandling . . . perpetuates and increases homelessness through enablement" and "The mission should no longer be to 'serve' the homeless community."

Cities hire Marbut to "research" their "homeless problem" and present his findings to the city council with recommendations for solution. Inevitably, he recommends building a one-stop-shop, come-as-you-are, twenty-four-hour emergency shelter like Safe Harbor—or what he calls a "transformational housing portal," typically run by the sheriff's department—and sweeping all of the area's unhoused people inside it. There, they find everything they need: housing assistance, medical care, drug rehabilitation—Safe Harbor even has Wi-Fi. While construction is underway, Marbut continues to get paid.

Safe Harbor is a jail-diversion program initially designed for those released from the county jail with nowhere to go and for those who were cycling through the system on municipal ordinances targeting homelessness. I visit the facility in July 2015. "Homeless get arrested, homeless go to jail, homeless get out of jail, then come back, and do the same thing over and over again," says the deputy leading me on a tour, who asks not to be named. They get arrested for trespassing and are hit with a fine or a notice to appear, which they inevitably don't fulfill. Next time a cop sees them, it's an automatic trip to jail. It's a burden on the taxpayers and a burden on the sheriff's office, he says. The services that go on inside a jail are expensive. So are the public defender, who will need to defend these people, and the state attorney, who will prosecute them. Meanwhile, none of those expenses help prevent the alleged crime. The idea is to take these people to Safe Harbor rather than to jail. It costs $13 a day to keep a person in the shelter—compared with $125 a day for the jail. Big savings for the sheriff.

My tour of Safe Harbor begins at the metal detector set up at the corner of the outdoor pod. Originally designed to hold 250 people, Safe Harbor now holds a daily average of 450, separated by gender into five housing areas, or pods. Four of the pods are inside, and one is outside. The outdoor pod is a slab of concrete sectioned off with a chain-link fence, two metal tents overhead, and a set of porta-potties. It's intended for those who behave badly or show up inebriated, though it's not unusual for people to spend the night here while waiting to get inside. As many as a hundred people have slept out here on a single night—and it rains a lot in Florida. The deputy, whom I'll call Jake, is warm and loquacious. He later describes himself as a "big softie." He explains that, while the shelter was originally intended to employ a large staff of caseworkers, the funding has since fallen away and they now employ only four, with two more employed by WestCare, a local mental health service organization.

"The guy with the bald head and the black shirt," Jake says, gesturing to a man conferring with a cluster of residents. "His name is Charlie. He's the case manager for the folks that are living outside here. He's employed by WestCare. His whole focus out here is the folks that have substance abuse—the great majority, I don't have any numbers on it, but I would say easily estimated at about ninety percent of our folks have mental illness. And ninety percent of those folks are treating that with substance abuse, or are unwilling to get treatment. So, those are the folks that don't want to follow the rules, so those are the folks that wind up out here."

Safe Harbor works on a reward/punishment model. The philosophy is that clients are rewarded for good behavior with increasing levels of comfort, the pinnacle of which is housing. Likewise, they're penalized for not working hard enough, coming back drunk, or behaving badly inside—they go back to the outdoor pod.

"We want to move them in as fast as we can," Jake says, "so Charlie's out here, and me and him are in communication every day. Who can we get moved inside? Who might not have come back sober last night, but if we move them inside, will they come back sober? A little carrot on the end of a stick to get people to do something right."

It occurs to me that Jake might have it all backward. As we pass the check-in area on our way inside, I remember something G.W. said after that breakfast a few weeks ago. We'd finished cleaning and gone outside for a cigarette. He pointed to a man I'd met earlier, JR, who'd had the same hip replaced twice, had diabetes, and walked with a cane.

"JR, they cut off three of his toes, but he's not sick enough to qualify for a disability check," he said. "He sleeps on the sidewalk every day, goes to the hospital every day so they can infuse him with antibiotics" after an infection hospitalized him for weeks. "Every day! Tell me you're not going to have something alcoholic to negotiate the sidewalk."

Many substance abuse experts agree with G.W. Substance abuse is both a cause and result of homelessness, and the majority of people abusing substances on the street are mentally ill, so treating substance abuse alone is inadequate; it must be tackled alongside homelessness simultaneously. As G.W. says, put them in a house and wrap services around them, and ensure that they can't lose their house no matter what. Only once people are safe and secure in their living arrangements can they focus on themselves. This approach is commonly known as "housing first." It proceeds from the idea that housing is a human right and not a privilege. For the last ten years or so, it has been widely acknowledged as the only effective way of ending homelessness.

Safe Harbor is stark: The hallway and floors are concrete. A central hallway connects the whole building. Rows of lockers line either side. People are busy painting the walls yellow as we move through. We pass an inmate from the jail employed by Safe Harbor. Inmates come over to do the cleaning and the chores that Safe Harbor doesn't encourage residents to do, Jake explains, because—he lowers his voice—sometimes people are lazy. They can ensure that an inmate is going to get up and do the work. Low-custody, nonproblematic, low-flight-risk guys doing six months or less. These men have never had drug arrests, and they have never been to the shelter before, so they don't know the ins and outs of it—they're unlikely to steal things or traffic drugs. They work for eight hours a day; then they walk back to the jail.

Jake stops in front of a large open area used for storage. An office stands on either side of the entryway; caseworkers share one, and twenty-five interns from the public defender's internship program share the other. The open area is where classes take place: work readiness, addiction recovery, and budgeting among them. Donations of clothing, bicycles, and food are received here. Clients can stay as long as they want so long as the staff sees that they're trying, Jake says. They also discourage clients from doing only temporary work and day labor. They want them to succeed, and they tell them to go if that's the only work they can get—but it's not going to get them off the streets and keep a roof over their heads. It's going to keep them in shelters.

"A lot of our folks show up, sometimes, they've been living in the woods or off the grid for the last ten, fifteen years," Jake says. "They have no work history, they have no ID, they have no social security card, they have no birth certificate. They walk in the door and they're a clean slate. They have to work very hard to get all those things; it's hard post–9/11 in this country. They're going to sit for six to maybe eight weeks." Only then can they send them out to find a job.

Three pods house men, and one houses women. Numbered areas taped along the concrete floor designate sleeping assignments—clients sleep on thin, hospital-blue, plastic-covered mats like the kind you see in correctional settings. This saves the county money and prevents bugs from spreading.

"At the same time, we don't want [clients] to be so comfortable that they don't want to leave," Jake says. "A mattress on a floor, it's not glamorous. But it's not a bus stop, it's not their car, and it's not jail."

Clients shower and shave in communal bathrooms. Plastic tables and chairs fill the center of the room, on the far side of which a door leads out to a fenced-in rec yard, open twenty-four hours a day for people to smoke a cigarette, clear their heads, "do what they got to do." These people have a lot on their minds. Safe Harbor sees a lot of people that are waiting on Social Security Disability Insurance approval, for instance. That process can take a long time, from two months to two years. "If they're unable to work and being advised by their lawyer not to work, it's tough to get them money," Jake says. The county has disability advocates—Safe Harbor encourages people who haven't yet secured a lawyer to go through the county because it's a lot faster. If a client has an external attorney, sometimes they can wait two years before getting approved. Still, even people going through the county are usually turned down the first few times they apply for disability. Once they have their meeting with a judge, it takes another sixty days to get a decision.

"So, what's their recourse in the meantime?" Jake says. "They bounce from shelter to shelter."

I ask why the process takes so long.

"Ask the federal government," he says. "I don't know."

I ask if these are people with legitimate disabilities.

"I don't want to say 'legitimate disability' because the judge is saying they don't have a legitimate disability," he says. "But, you know, they present as disabled. They've got a walker, they've got medical records to show they're disabled."

We continue down the hallway. In a separate pod, veterans and people who work or go to school full-time have the privilege of sleeping on real beds whenever they want to—that's the third tier of the program, after sleeping outside and sleeping on a mat on the concrete floor.

The last pod is the women's. There are more than four men for every one woman at Safe Harbor, roughly representative of the homeless population as a whole. Men and women are not allowed in each other's pods. Jake hollers as we step inside: "Male coming in, ladies."

A woman sits alone at one of the tables doing a crossword.

"Twenty-four/seven, doesn't matter what time of day it is, there's always a woman sitting here," he says. "Not because we tell them to—they're very protective over their area. A lot of them are victims of different things."

On our way back to the intake area, Jake stops by his office. A daily count of Safe Harbor's veterans hangs near the door. The number hovers between thirty-five and forty on any given day.

"We don't do follow-up case management," Jake says, handing me a business card. "So, once somebody walks out the door and has other resources to follow, unless they come back, we don't know about them."

I ask if that's because they don't always know how to reach them.

"I don't have the funds, I don't have the availability. No staff, no money for it. It's not something we're interested in. We're primarily here to keep people out of jail."

I ask him how well he gets to know the people who come here.

"Some of them, we'll see them for two weeks, they save up enough money to put down that first, last, whatever, they're out the door, we never see them again," he says, leaving me at the outdoor pod. "I've been at Safe Harbor for two years. Some of them I've known since I started at the sheriff's office thirteen years ago. They're in and out of jail all the time. They're starting over in life for the seventeenth time. Some folks—when we're talking about the level of mental illness, substance abuse issues that these folks have—are in and out. They're trying over again and again."

The following Saturday, I arrive again at G.W.'s breakfast at eight o'clock. Pastor Tom Snapp is sitting outside the elevator in the fellowship hall. Snapp is part of Celebrate Outreach, the consortium of churches that hosts the breakfast and dinner, along with G.W. and fourteen other local faith workers. Snapp is frail, in the early stages of chronic obstructive pulmonary disease. He sits in the hall and counts the people who enter, and gives counsel while the others serve breakfast. People know to come up to him for guidance. This morning, I stand by while he talks to a man whose wife has a developmental disability. "I married them," he tells me later. "Right here on the stage."

The kitchen is lime green and outfitted with two refrigerators, a large restaurant-style grill, a row of industrial sinks, and an island counter. G.W. is goofing around with a couple of volunteers. One stands at the grill frying potatoes. One is cutting corn bread.

Another is doing the dishes. Two serving trays of eggs sit atop a plastic rolling cart at the edge of the grill, one with ham and cheese, the other plain. A large bowl of cut fruit sits covered in plastic on the island next to a serving tray of grits. A pot of coffee collects condensation near the door. I watch as the volunteers finish preparations and wheel the food out to the plastic tables lined up under the windows. We circle up again for a prayer.

G.W. begins. "If you pray for something and you believe in it, and act towards it, you can get that thing," he says. "This breakfast ain't nowhere near being over. Lack of money ain't never stopped me from doing nothing."

People look at one another.

"If we have to do the grits like we do the eggs now, we can do that. I got stuff. And I can make pancakes from scratch. We might not have eggs every week, but we'll just have faith. We don't need eggs for breakfast every week," he says.

He steps back into the circle. We all take hands. Tom Snapp speaks, "Lord God, we thank you that we're here today. We ask you to send funds for this breakfast, that we can continue. We thank you for those who have come to serve, and those who have come to prepare, and those who've come to eat. Bless us all with your grace and presence and love. We pray in Jesus's name. Amen."

We wear plastic gloves and greet each person. Again I serve grits. Some people ask for the grits on top of their eggs. Some crack jokes. Some prefer not to make eye contact. I recognize a woman with paint around her mouth from the last Saturday, and a man who wears his bandana down over his eyes. Couples move through the line together, bickering or talking sweetly, holding plates for each other. About 150 people attend the breakfast each week, and more toward the end of the month before they get their government checks. It costs $240 to feed them all, some of which comes from grants, but most of which came from the budget of Trinity Lutheran Church and the congregation doesn't want to pay anymore.

"It's been going for six years," says Tom Snapp, counting as people file out of the elevator. "G.W. started it. We've got enough for one more week. I made an appeal to the congregation several months ago, and they gave a couple thousand dollars. But you know, at two hundred forty dollars a week, it goes quickly, and I don't think the congregation's in a mood right now to want to shell out more and more."

We get to the end of the line and start serving seconds. As the food runs out, volunteers carry away the trays. A few people come back with empty pastry containers and ask for food to go, for friends who couldn't make it this morning.

"Everyone consistently says it's the best breakfast in town," G.W. tells me later, smoking out on the steps. It rained during the breakfast and the sidewalks are heating up, making the air kind of balmy. People mill around the church entrance, sleepy after filling their bellies. "St. Vincent was never a breakfast place. They'll give you a frozen bagel and a hunk of cream cheese. Watery coffee. And just call it breakfast. This breakfast is made with love and care. The cooks are real good, you know? I'm kick-ass. Now I train other people to do it because I don't believe in top-down solutions. I don't believe in solutions that are brought to you by a committee that doesn't have any skin in the game."

G.W.'s most recent episode of homelessness began in St. Petersburg in 1998. He was living in a mother-in-law apartment behind a house, and it burned down. He was cooking then, in various beachside restaurants, and doing the books for his landlord, who managed some small companies. He lost everything he owned in the fire. He thought he could get another place immediately—then he lost his job. Then someone stole his belongings while he was sleeping. He couldn't clean himself, nor present himself as a chef. He lived on the streets for the next eight years.

"When I was homeless," he says, "the people who were lying on either side of me were coming to shake the bush, to get me out of the bush, so that I could take a birdbath and go cook for people."

He's interrupted by his phone ringing. On the other end, a woman he's known for years asks him for money. She secured supportive housing through Boley, a local agency that serves people with mental illness. But she's brought another family to live with her—eight additional people, along with her and her daughter—and her government check has run out. G.W. tells her he'll meet her later.

"I'm just scared they're going to get kicked out," he says, hanging up. He puts out his cigarette, collecting his thoughts. "The power of transformation is the power that people have already inside themselves that they're not aware of, that hasn't been exploited," he says. "If you don't identify power in somebody and give them power, then you haven't transformed power. I feed people, but I usually call it food sharing, or food distribution. Because when you 'feed people,' they sound like animals in a zoo. The only responsibility that they have is to accept the food, eat it, and chew. I try to make people see that we're feeding each other."

So the breakfast helps him, too, I say.

"I'd take a bullet for every single one of them," he says. "I wasn't religious when I was on the street. I was just trying to be a good guy. Just trying to be a factotum—the barber of Seville, you know? Invent a constituency of people and deal with them every day, their pains and their dealings. I had all this time, and time is a killer. Time on the street? What are you gonna do on the street? Ed the Mop, he comes here every Saturday and he mops the floor. He stayed in there and cleaned up as long as he could today, and now he's on the street. So what is he gonna do? Gonna wait for supper somewhere, and just idle away his time."

I ask him what it's like trying to find a job when you're living outside.

"It's futile, mostly. Nobody's going to hire a homeless person. A lot of people, when they get inside, they just erase the evidence that they were homeless, but I didn't do that. It's hard to have a society when you get out of homelessness. Who are your friends going to be? I turned around and walked back to the people."

A woman has joined us on the steps. "Can I say something, a thing you said that was true?" she says. "You're right, when people are homeless and they get off the street, a lot of times they do separate themselves. Half the time, they do so because they don't want to be reminded of the hardship. Of sleeping in the rain."

G.W. nods. "I really believe that three or four years down the road, homelessness will end," he says. "I really believe that God brought me to this point, and to this particular place, so that I could contribute toward the end of homelessness."

I ask him how homelessness is going to end.

"We have the money to end homelessness now," he says. "We have to provide affordable housing. And if we are successful in the argument of supplying affordable housing, and making sure that everybody has a means to get it, homelessness will end. But we have to come up with a story. We have to have a story that resonates with people who will say, 'Yeah, man, we're the richest country in the world. Why are people sleeping on the street?' The enemy has sown weeds in the wheat saying, 'They want to, or they're drunks.' Even if they're drunk, they shouldn't be sleeping on the street. Most drunks live inside."

In 2014, three years after he initiated the program, the City of St. Petersburg called Marbut back for a follow-up evaluation. The homeless problem was creeping back in; people were again congregating in Williams Park, and downtown business owners were complaining of people loitering outside their storefronts. While Marbut claimed the city's nighttime population of homeless people had decreased, the county's point-in-time count showed that the number of people experiencing homelessness in Pinellas County hadn't changed—it had just moved north from St. Petersburg into the area around Safe Harbor, much to the dismay of High Point residents nearby. The problem overall had gotten worse: That January, Homeless Leadership Board volunteers counted 5,887 total homeless individuals in Pinellas County—exactly what they'd counted in 2011 when they built the shelter. From 720 unsheltered people that year, they were now up to 1,178. Over 2,500 of those counted were children. Half of those counted lived in St. Petersburg.

"It's obvious to anyone paying even a little attention that Williams Park has gotten significantly worse," council member Karl Nurse told the *Tampa Bay Times*. "We've got to step it up."

Marbut made nine visits to St. Petersburg and delivered his findings to new mayor Rick Kriseman and the city council on June 5. "This is kind of a good news/bad news/good news situation," he began. New daytime "hot spots" were popping up around the city, and about forty-five to sixty-five people were floating between them, mostly between the hours of six and eleven in the morning—that's after shelters kicked them out, and before the new day program at St. Vincent de Paul, the city's largest homeless shelter, let them in. Marbut noted that St. Petersburg police didn't seem to be engaging much with homeless people anymore. "That's a real concern," he said.

He made six recommendations, the first of which was recommitting the police department. The others included deciding whether street outreach teams should focus on individuals or families, because they weren't doing a good job splitting their efforts between the two; starting a whole new program to service homeless families alone; putting more money into St. Vincent's day program, to open it earlier; putting more money into Safe Harbor; and reinstating the sheriff's Chronically Homeless Jail Diversion and Intervention program.

"Does Robert Marbut have any money in his pocket?" Sheriff Bob Gualtieri asked the *Tampa Bay Times*, noting that the Chronically Homeless Jail Diversion and Intervention program was very expensive. "He breezes in and out of here and calls for things, but who is going to follow through?"

The ninety-day Chronically Homeless Jail Diversion and Intervention pilot program had come to an end the previous October. Through the program, 155 of the county's chronically homeless individuals, including its twenty-two "most recalcitrant," were diverted to an effort intended to steer them into addiction-recovery programs or into transitional housing if they were already sober. Under the Chronically Homeless Jail Diversion and Intervention program, individuals arrested for minor infractions, such as an open-container charge, were held in solitary confinement for an indefinite period of time instead of being released within the usual twenty-four hours. During this time they met daily with Safe Harbor and Pinellas County public defender personnel, who encouraged them to transfer out of the jail and into Safe Harbor. At Safe Harbor, the sheriff said, people struggling with addiction could be assigned a bed at a treatment facility—which would help them get clean, though not help them find housing. When asked by the *St. Petersburg Tribune* how long he was legally allowed to hold these individuals, Gualtieri responded, "We'll have to deal with it as we go. You are asking for an answer to an unanswerable question."

Only not everyone wanted to go to Safe Harbor, even when faced with the threat of more jail time. Out of 155 program "participants," only 93 individuals opted to go to Safe Harbor. Safe Harbor isn't that different from jail—and at least the jail is air-conditioned. It's hot in the outdoor pod, and it rains. In jail, they still get three meals a day and a plastic mat, and nobody tries to force them into treatment. So why bother with the shelter?

And because Safe Harbor doesn't do follow-up casework, there was no way to know whether or not the Chronically Homeless Jail Diversion and Intervention program actually helped get people off the streets for good. Everyone seemed to agree it did, but the best analyses were anecdotal: "We are not seeing them as much as before," Major Dede Carron, who ran the homeless outreach program for the St. Petersburg Police Department, told the *Tribune*. "They are staying in jail longer and they seem to disappear after, so we are presuming they are going into treatment."

The choice of being taken to Safe Harbor instead of the jail was already available to homeless people arrested on petty ordinances before the Chronically Homeless Jail Diversion and Intervention program had been instated. And increasing numbers of homeless people familiar with the system, even those who weren't part of the Chronically Homeless Jail Diversion and Intervention program, were opting to go to jail instead of going to the shelter. Consequently, the number of people identified as homeless when booked into jail had gone up since just a year after Safe Harbor opened, from 17.6 percent in 2012 to 23.5 percent in 2013. But closing it wasn't an option, said the sheriff. Enough people still opted to go there over jail—and jail costs the sheriff's department ten times more.

The sheriff's department shoulders half the expense for Safe Harbor. Of the shelter's $2.2 million annual cost, $1.6 million comes out of the sheriff's budget. The rest comes out of the county, and those costs are offset by donations from various cities. Clearwater chips in $100,000, as it sends the second-highest number of homeless citizens to the shelter. Largo has never contributed, citing increased stress on its emergency services, since the area around Safe Harbor falls within its jurisdiction. Pinellas Park, Dunedin, and the beach communities pitch in what they can. St. Petersburg's share is the largest: upon Marbut's most recent suggestion, it tacked another $50,000 onto its standing $100,000 contribution. It also increased its police presence downtown: officers began paying special attention to Williams Park and other Marbut-identified "hot spots" and stepping up their arrests for petty ordinances.

All seemed quiet for the next year or so. Then, in the summer of 2015, the *Tampa Bay Times* reported that in the last two years a program run by the City of St. Petersburg had bused close to a thousand homeless people, some of them addicted to drugs or mentally ill, across the country. But after busing these people to who knows where, the city didn't make any follow-up phone calls—there was no way to know for sure where they ended up. A street outreach officer told the *Times*, "I'm not a caseworker," and other officials voiced similar opinions. The *Times* reproached the city for its lack of accountability and "not in my backyard" attitude—but admitted the program was very popular among those being bused. According to the article at least one rider confessed that getting his ticket out of town was no less than a matter of life and death.

The day before Marbut's meeting with the city council, Michelle Obama launched the Mayors Challenge to End Veteran Homelessness, an initiative that enlisted mayors, governors, and county and community leaders nationwide in the effort to house every homeless veteran by the end of 2015. The US Department of Veterans Affairs had first announced the mission to end veteran homelessness in 2009—now the White House was calling on localities to get serious. The First Lady told the US Conference of Mayors, in Washington, DC, that roughly 58,000 veterans were experiencing homelessness in America at the time. Veteran homelessness had dropped by 24 percent in the last three years with the institution of Supportive Services for Veteran Families, a program keeping low-income veteran families from falling into homelessness, and the strengthening of HUD-VASH, which gives housing vouchers to homeless veterans. But federal programs could only do so much, Obama said. In some of the nation's largest metropolitan areas, there may be as few as a few hundred homeless veterans. Communities know where to find those veterans, and know their needs, so should know how to house them.

In January 2015, Mayor Kriseman of St. Petersburg took up the First Lady's challenge. The point-in-time count that month had revealed an increase in the overall number of homeless people in Pinellas County—up from 5,887 the previous year to 6,853, including 621 veterans in Pinellas County alone, of the almost 3,000 total homeless veterans in the tri-county area. Kriseman initiated the Pinellas County Task Force to End Veteran Homelessness, a countywide effort among cities and service-providing agencies headed by St. Vincent de Paul, St. Petersburg's largest homeless shelter, to identify, find, and house every veteran.

When I talk with St. Vincent de Paul's executive director, Michael Raposa, about the project in January 2016, he tells me the tri-county area is now home to less than two hundred homeless veterans. "To say that the housing doesn't exist is an excuse," he says. We sit alone at a boardroom table in his office, overlooking a roomful of caseworkers chattering on their phones. "We've got a database of four hundred fifty-four, last year, landlords that we used. It's all private stock. Sometimes clients have a VASH voucher, or a Section 8 voucher, but most of the time they don't. It's only for about twenty-five percent that we're using subsidized housing. Most of the time, they can do it on their own."

St. Vincent de Paul works with clients to find suitable locations and housing within their financial means. Most of the units are in smaller complexes of eight or less, as it's easier to form relationships with those landlords. Landlords in smaller complexes tend to have more freedom in deciding whether to overlook a criminal record, for instance. The goal for housing clients is thirty days after first contact. Once they're housed, their caseworkers connect them with services and disability insurance, if they need it. They help them with budgeting. They even help them budget for illegal drugs.

"We help them see, 'You have a thousand dollars coming in. First thing that has to go out is your rent and utilities. You've got four hundred dollars to play with a month, so one hundred a week for four weeks of discretionary income. You're going to have to eat off that unless you go to the soup kitchen. What's your drug habit?'"

This seems very forward-thinking. But, I ask, doesn't addiction tend to get more expensive over time?

"There's a difference between drug use and drug abuse," Raposa says. For those using it to treat mental illness, they work with them. Adherence to their prescribed medications is critical. "But we tell them, point-blank, 'We don't care what you do. We just can't have this housing fail.'"

I ask him what the biggest challenge facing formerly homeless people is once they get inside.

"Loneliness," he says. "All the social workers are surprised by this. I'm not." People in poverty are motivated by relations, he explains. People in the middle class are motivated by achievement. "You've probably got a college degree."

I say that I do.

"The vast majority of homeless people are generational, meaning that their parents had tenuous housing, and their grandparents had tenuous housing," he says. "And we're also seeing that the age difference, in some situations, between those generations, is only twenty years. We've got grandparents at thirty-two, because they had a kid when they were sixteen, and their kid had a kid when they were sixteen. Now they're a grandmother, and taking care of a grandchild."

Earlier, in the waiting room, I'd sat next to a young mother with her child on her lap. She was tattooed and pierced, with ripped jeans and ironic socks, reminiscent of the way I dressed in high school. She couldn't have been older than nineteen or twenty. Her son played quietly on her phone. She kissed him on the back of his head and the tips of his ears. She held his hand when a woman called them up.

Raposa says, "We're trying to tweak a system that actually has a deeper understanding of poverty, and can work within the framework of those issues in a nonjudgmental way, and actually lift people up as opposed to standing on our mighty, middle-class high horse, thinking, 'You should want this horse because I have one.' Truly, that just doesn't work.

"Military recruiters target low-income populations. It's not that much of a stretch to realize that these guys didn't have a lot of ace skills going in, they went in and served, they were released in an economic environment where college graduates are having a hard time finding a job. It's low-level jobs, and you're living in a community where you need to be making $14.50 an hour just to survive. The jobs that they're getting are minimum-wage—$8.05 an hour in Florida. They deserve more. On top of the psychological stuff they're dealing with."

St. Vincent de Paul serves the entire homeless population, not just veterans. On the way in, I had passed several blocks of people sitting on the sidewalk or lying on the grass, or resting inside St. Vincent's caged-in courtyard beneath the freeway overpass. I ask Raposa what needs to happen in order for the organization to bring the housing-first services they're bringing to veterans to the nonveteran population as well. There is almost no rapid-rehousing money for nonveterans in Pinellas, he says, but they're working to fix that. The Homeless Leadership Board has just submitted an application to HUD for a half-million dollars, though that won't even scratch the surface.

"Those numbers are off the charts," he says. "Pinellas is about six years or seven years behind where we are with veterans. It's going to take a huge investment to catch it up." But what they're doing for veterans now is clearing the path. They're demonstrating performance, which makes it easier for less progressive city and county officials to see the efficacy of the housing-first philosophy. In fact, St. Vincent's has already been talking with the City of St. Petersburg about funding a three-year $2.25 million initiative to house every person sleeping in the area around St. Vincent de Paul, he says. It's a lot of money, but the city is considering it.

The consensus at the next July breakfast is that the powdered eggs look like dog diarrhea. No one had read the can closely: they're filler eggs for baking, not eggs for eating alone. We stand around the island counter looking at the pan. Fresh from the oven, the top layer has congealed into a matte brown mass. Rita, a volunteer, pushes it aside with a spoon. "Father Lord Jesus in heaven, we ask you to beautify these eggs so they can work in your service for people," she says.

Something is wrong with the fan in the kitchen. Steam rises up off the grill and from the dishes in the sink, sticking to the sides of our faces and saturating all of the towels. G.W. steps out to catch his breath in the fellowship hall. I follow him to a far back table near the stage and ask him what's bothering him aside from the eggs.

"I've made mistakes this week," he says. "The eggs, and the expenses . . ."

I wait for him to finish.

"I wasted some of the money," he finally admits. "I ain't gonna tell them that. And in my personal life . . . I relapsed."

G.W. has never been clean in the twelve-step sense. He loves alcohol. He takes pain pills. But he's made a new friend who smokes crack. They like each other, but they can't last this way. They've been smoking together for a few weeks. He has to stop.

"God is telling me I'm already free," he says. "These are the times when God is most evident. You ask God to walk you through crossroads."

He takes my hand.

In 2009, shortly after I met him, G.W. was hospitalized with liver cancer. We didn't know each other well, but something sent me to the hospital with a bag full of magazines to visit him. When I entered the room I was surprised to see him in a gown, this man I usually saw in a button-down shirt. Something about the way he's holding himself reminds me of this now.

At sixty, G.W. isn't a healthy man. His walk is more of a shuffle, and he often struggles to breathe. He leans against whatever is nearest him when he stands. When he climbs in and out of my car, I frequently ask him if he's okay. I ask him now if he's going to seek treatment.

"My needs lay where I am," he says. "I will find a way to stop. You know why? Because I told you about it."

He'll move out of his apartment, he says, away from his friends who are using. He'll get his head straight, get back on track with the things he's doing, start reading again, start thinking the way he needs to be thinking.

"I'm of the opinion that God doesn't give a damn what you do," he says. "Jesus said it's not what goes into man that defines him, it's what comes out."

I ask him if the person he's smoking crack with goes to church.

"No." He laughs. "She's astounded I'm a minister."

He was up late last night and is very tired. He makes it through breakfast and gets into my car afterward. I drop him off at his house and he struggles up the walkway, breathing heavily. I say a prayer as I pull away, watching him in my rearview mirror.

By the end of *Easy Street*, G.W. finds a way inside. He's gotten a job and is making money. For the first few weeks, he still sleeps in the park, washing up in the public restrooms in the morning and stashing his bags at someone's house during the day. He takes the bus an hour and a half to work at a travel agency selling vacations over the phone. He makes commission. None of his colleagues know he's homeless. He lives in fear of someone finding out, and in fear of someone stealing his money while he sleeps. Weeks into the gig, he's able to rent a room at the Kelly Hotel.

"I didn't really want to get into a hotel room," he says as he walks across the stark lobby, reading a newspaper. "I really tried hard to find an apartment, but I couldn't. I looked everywhere."

We cut to him sitting at a small wooden table next to a refrigerator. There's a newspaper, a pack of Seneca cigarettes, a lighter, an ashtray, and a pair of sunglasses within reach. He sips a forty-ounce bottle of Natural Ice.

"I pay one hundred seventy-five dollars a week for this room, which is probably a little steep," he says. "There's no cushion in a hotel room, you know, with this job, and with a hotel room and having to pay week by week, anything could happen. I could get fired. I could get sick. I could lose my voice and not be able to talk on the phone. Anything could happen. The first step was getting a job, and I got that, and I'm pleased with that. The next step is getting an apartment and getting a place of my own."

He microwaves a prewrapped sandwich he bought at the convenience store around the corner. He sits on the bed in his socks, eating his sandwich, with the newspaper spread next to him. The comforter is a loud floral pattern, in jarring contrast to the room's ascetic interior.

The next time we see him, he's in his office. He sits at his desk in a small gray cubicle. He wears a clean white button-down shirt and khaki pants. The office is decorated in a tropical theme, with paper palm trees and Christmas lights. Above his desk, two laminated signs read SINCERITY & ENTHUSIASMS and EMPATHY & CONVICTION. His desk is outfitted with a mirror, so we can see his face while looking over his shoulder, as well as several people passing behind him.

"I started not selling," says his voice-over. "And it's like, I knew that you go through cycles of not selling." His boss told him if he didn't sell two vacations by the end of that day, then he'd be fired. "And I said, 'Well, how about this: I'll make three more sales, and then I'll tell you to kiss my ass and quit,'" G.W. says. He shuffles some papers.

That day, he collapses in the office. Paramedics rush him to the hospital in Largo. They sedate him on June 29. When he wakes up, it's July 15.

We cut to him sitting on a bench in Williams Park. He is thinner, and tired. By now, he's lost his job and his room at the Kelly Hotel, and is sleeping on the floor of the Dream Center, a faith-based organization serving the homeless. He stands, turns, and lifts the bottom of his shirt. A scar runs diagonally across his back, from his shoulder to the bottom of his rib cage. It wraps around to the front of his body, where it meets another scar, horizontally, in a triangle.

"I still don't know what the operations were for," he says. He pulls his shirt back down.

"It seemed like they had me in a basement," he says, describing the hospital to me later. "It was more like a morgue, in a nurse's room. They were racist and mean. I was hearing conversations about them stealing my drugs. I dreamed a lot about my brother and my sister, who was alive at that time. I thought I had those dreams every night, but it was an ongoing dream because I didn't know day from night. I wasn't aware of temporal reality."

Prior to falling ill, G.W. had been hanging out at the Dream Center, helping them cook morning meals when he wasn't at the travel agency. He'd learned how to cook in New Orleans, while he was a merchant marine. He cooked for hundreds of people at a time, big meals, for five years at sea. After cooking at the Dream Center, he'd go over and hear the church services. "I just incorporated that in my society," he told me. "Then I got sick."

His friend Tom, who worked at the Dream Center, heard that G.W. was sick. He went down to the hospital to advocate for him. They wheeled G.W. into surgery. His lungs were filled with bile and rot; they opened them up and cleaned them out. He flat-lined twice, and twice, at Tom's insistence, they kept on working. Meanwhile, G.W. dreamed—

"I was riding in a car with people I knew from New Orleans," he told me. They told him to put his clothes in the trunk and he obeyed—he hadn't realized he was naked. He walked through a door and found himself in a courtroom. "They were saying, 'Well, what has he done?'" he said. "I was trying to talk, and it was kind of like I was a child pulling on people's pant legs. They were ignoring me, and they were talking about me, telling stories."

The prosecutor said, "He's known this story since he was a child, okay? He's known this. Sometimes he has adopted it, sometimes he's not." They decided he wasn't ready.

He leaned against a wall and it spun around, and he found himself in a field.

People were celebrating and eating something like chicken. A black guy came down and everyone started clapping. G.W. asked who he was, and the people said he's the one who invented this: "He said that if he were ever famous or rich, he'd share his profits with his employees. And he's rich and he's famous, and so he did. He shared. He shared it with everybody."

Just then, somebody called G.W. and said his job was ready.

"I left the field and I went to this big, huge kitchen. It had everything you need—it had a steamer, a kettle, a tilt kettle, and all of the kitchen tools. They said, 'This is where you're going to be working if you get the job.' I said, 'Gee, I hope I get this job!' and everybody started laughing. They said, 'Kid, you don't want this job.'"

They called him back to the garden. They said they were going to give G.W. another chance. There were clothes, so he put on his clothes again. A huge black man was sitting there. "I said, 'Where am I?' He said, 'Young man, don't go in there, because the court is going to be tough. These motherfuckers ain't bullshitting.'"

Then he woke up.

He got out of the hospital three days later and slept on the floor of the Dream Center. He received a $300 rent voucher for his disability—and somebody accepted it. "And what do you know?" he said. "I'm not homeless anymore."

The person who accepted the voucher was an elderly woman who lived alone with her cats. When she heard G.W.'s story, she cleared out her guest bedroom so he could live with her. *Easy Street* ends with this segment: G.W. stands outside his new house. A rainbow wind sock in the shape of an old woman hangs over a set of moss green stairs leading up to a latticed porch, which the old woman calls her "cattery." He wears a long-sleeved, gray-and-white-striped button-down shirt and sunglasses.

"This is 672 Preston Avenue South," he says into the camera. "This is where I reside."

The old woman crosses the yard with a walker.

"This is the landlord here."

"Just walking my dog," she says.

We cut to them sitting together at a marble table in the backyard.

"Here we are in this country," she says. "We go to the moon and all this fancy stuff. And we can't even find a bed for people. That's pathetic. It is shameful."

The filmmakers ask her how she gets along with G.W.

"Oh, so-so." She laughs. "He goes his way and I go my way."

We cut to G.W. in the kitchen, frying something on the stove.

"His cooking style is a little different than mine, but we get on some things like the beans and rice," she says. "Now that's good."

We follow the old woman into the house, through the cluttered living room that still holds the contents of what is now G.W.'s room. His room is painted pink. She stands before a desk strewn with hygienic items, vitamin and medication bottles, soda bottles, and papers. She points to a clothing rack behind a foregrounded bookshelf holding books and a stack of newspapers. The camera rotates, and we see, hanging between two windows, a wooden cross. It continues rotating, and we see G.W.'s bed, which looks very comfortable, with a pink coverlet and fluffy pillows.

"So, this is it," she says, "and it looks out on the backyard."

For months after G.W. left the hospital, he took the bus three hours every day to get an antibiotic shot. "All I had to do was get a shot so that I could go to sleep," he tells me. "I would pray all the time." A friend from the Dream Center gave him an audiobook of the Bible. He listened to it on the rides to and from the hospital. "I had always been neutral, and I always tried to do the right thing, since I'd come out of prison," he says. "I just started trying to see what I could do for people when I got well. It's like God told me he wanted more out of me than what he got." So, he became a minister. "It's constraining sometimes," he admits.

I ask him why.

"I want things that I know I cannot do. But Paul says, for men like me, all things are lawful—but are they beneficial? I don't get freaked out about people swearing, having dope, watching porn, and whatever, because those are temporal things—things of this world. It's: How do you *act*? How do you feel? How do you really feel about people? That's what you have to fix: how you really feel about people."

He was ordained by Bruce Wright, his friend and another local minister and homeless advocate. A few years later, he founded Missio Dei in the fellowship hall of a church near downtown. It was the natural next step, he said, like getting ordained. He later realized he'd always leaned toward religion. But he hated dogma.

"I've been writing since I was seventeen, in prison," he says. "I used to write every day and I read a thousand books—and I'm supposed to stop [writing] because of *your* religion? I'm supposed to stop because of *your* morality, and stop honestly saying the things I see and feel? I mean, organized religion sucks. It really does. But it all boils down to that six-thousand-year-old question: What are you as a human being? How are you a reflection of God?

"You're supposed to fuck up. You're a human being. That's part of your freedom. But you should get up on the horse and try to ride. You shouldn't hurt anybody. You should save as many people as you can from adverse situations. That's the way I feel about it. That's why I do what I do without reward or publicity. That's why I became a minister. And I think I've told that story three times in my life."

In December 2015, the *St. Petersburg Tribune* reported Celebrate Outreach, the consortium of churches that oversees G.W.'s breakfast, was joining the effort to house St. Petersburg's homeless veterans. Homeless advocate George Bolden was leading the way: his Tiny Homes Project would build clusters of tiny houses under five hundred square feet apiece, complete with amenities. Each home would be designed through collaborations between architecture students from the University of South Florida and the veterans who would live in the houses and help build them. Once inside, they'd be given a case manager and would be required to take courses in homeownership, maintenance, and financial management. Celebrate Outreach was now scouting for land. One volunteer had offered up a vacant lot big enough for four houses, but Bolden was also looking for one big enough to accommodate a "village" of ten, which they'd call Eden Village. There are similar projects in the works in Oregon, Washington, Wisconsin, Georgia, and Alabama.

In his book *Tent City Urbanism: From Self-Organized Camps to Tiny House Villages*, urban planner Andrew Heben combines the horizontal democracy of homeless "tent cities" with the rise of the middle-class "tiny house movement." He attributes the rise in homelessness in the United States over the last seventy years to our inflated standards of living, in particular among the middle class. Compared to 1950, the average American now requires three times the amount of space: then, an average family home was 983 square feet and housed 3.38 people. By 2012, the average home was 2,500 square feet and contained only 2.55 people. These numbers have proven unsustainable. The 2012 census found that 10.1 percent of American homes—over thirteen million units—were sitting vacant.

While the mainstream conception of tiny houses, as featured on HGTV's *Tiny House Hunters* and FYI's *Tiny House Nation*, depicts the tiny structures as oversized dollhouses for the affluent, Heben's interest in them sprang from a desire to improve the physical infrastructure of self-organized homeless camps like the ones he was studying while maintaining their existing balance of privacy and social interactions. In his book, he looks at several examples of unsanctioned and sanctioned tent cities, as well as three tiny house villages operating under different degrees of government regulation. Among the tent cities he examined was St. Petersburg's own Pinellas Hope, which came about in response to backlash from the 2007 tent slashing, and about which Heben says, "my visit to the site . . . left me skeptical." The tiny house villages included Eugene, Oregon's Opportunity Village, which he helped establish.

Opportunity Village grew out of the closure of Eugene's Occupy encampment, which in 2011 attracted over one hundred otherwise homeless individuals. The closure instigated public desire to confront homelessness in the community, and in response Eugene's mayor formed a task force to find "new and innovative solutions." After months of advocacy and planning by Heben and his team, the first of Opportunity Village's structures was erected in August

2013, and in May 2014 the village was complete. In addition to thirty houses, it includes a gathering yurt, common kitchen, front office, tool shed, and a bathhouse with flush toilets, a shower, and a laundry room.

Heben attributes the village's success primarily to its grassroots style of governance, which balances the informal with the formal. It is self-managed but partners with a nonprofit organization and its board, which works with the City of Eugene to ensure the village meets municipal regulations. This informal yet intentional structure engages the community in the village's construction and operation, according to Heben. The project is enabled entirely by donations from private donors, businesses, and organizations. Altogether, subtracting in-kind donations, Opportunity Village was erected for approximately $100,000.

The support from the City of St. Petersburg for the Tiny Homes Project was overwhelming and seemed to signal a seismic shift in local perspectives toward the homeless community. Councilwoman Amy Foster, who had recently joined the Homeless Leadership Board, informed the *Tribune* in August that the city's new strategy for addressing homelessness was to provide permanent housing first rather than services. "It's a hard sell in the community, because people wonder, why would we give somebody a home if they're drug addicts or alcoholics?" she said, but "without housing, people can't focus on themselves." The city provided Celebrate Outreach with a list of city-owned properties for sale. The Tiny Homes Project was underway.

In the early weeks of 2016, the temperature drops and the city opens its cold-night shelters. G.W. and I go down to the church parking lot across the street from Williams Park and make sure people get into vans going to the best shelters. One, a substance abuse rehab facility, is called My Place in Recovery. Rita sits on a bench near a low brick wall guarding a Vitaminwater bottle filled with rum. "Fuck My Place in Recovery," she tells me. "I ain't recovering. You can write that down."

The sun sets, and as the chill settles in, a rumor circulates that the Starbucks in the Sundial complex is no longer giving out free cups of hot water. A pickup truck with a bed full of blankets and bottles of water shows up, empties itself, leaves. Ed the Mop waits along the edge of the parking lot with a rolling laundry basket containing his belongings. One man helps another out of a wheelchair and into the first van as it arrives, loading him in through the back double doors. When the third van leaves and we are left alone, G.W. points to a place at the foot of the low brick wall, behind a concrete bench.

"That's where I used to sleep," he says.

We go to the Emerald Bar and drink beer and hard liquor, and smoke cigarette after cigarette. Less than a month earlier, the week of Christmas, G.W. had a heart attack. For days, his breathing had been labored, and friends at the church had grown concerned. Rita saw him coming up the stairs one morning and knew in an instant that he needed to go to the hospital. There, a doctor told him he had the body of an eighty-year-old man, though he was sixty and looked forty. He asked G.W. point-blank if he was the devil. G.W. laughed. "I don't even like the devil," he said.

They stuck a camera inside his chest to see if he needed a stent. He didn't. He was told again to lose weight and quit smoking. He's going to quit, he reassures me.

He tells me he's been thinking about his novel. There are at least another hundred pages he needs to write. He has some people living with him at the moment, but as soon as they're gone, he'll get back into it. I'd met them when I picked him up that evening. His apartment was cluttered with their stuff, and it was hard to move around; it smelled like four bodies living in a place meant for one.

G.W. admits that it's hard to find time to be alone. The time that he does have, he spends preparing for the next weekend's meals. Somehow, money has materialized for the breakfast, but there are whispers that more will not be forthcoming. G.W.'s been laid off from his job at Southern Legal. Aside from his government assistance, he has no other income. Missio Dei pays the rent on his apartment. He isn't supposed to have others living there.

I tell him that I feel his novel is of urgent concern. He agrees.

The day of the January 2016 point-in-time homeless count, it pours. I arrive at my deployment center and find a cluster of people standing in a conference room reviewing the survey questions one at a time. I'm given a chartreuse T-shirt identifying me as a volunteer, a clipboard with twenty surveys, and a pencil. I climb into a silver sedan with my team leader and two other volunteers: an economics student from Eckerd College and a woman who works at the health department downtown.

Our first stop is Daystar Life Center, a drop-in agency whose stated mission is to "provide the basic necessities of life to our neighbors in need." People stand under the porch overhang, staying out of the rain. A man leaning against his bike says they've already talked to our group this morning. We proceed to the waiting area. A young woman sits watching TV, and a man sleeps in a chair against the wall. People talk at the check-in counter. Back outside, a man asks me if it's true that I have a stack of one-day unlimited bus passes in my pocket to give out as incentives for completing the survey. I say that it is.

I ask him for his first and last name and social security number—the last four digits are enough. His date of birth. How he identifies his gender. Whether he is Hispanic or Latino. His race and, if he is biracial, what he considers his primary race. If he is single, married, divorced, or widowed. If he's ever served in the US Armed Forces. If he has, I thank him for his service.

I ask if he lives with others who are homeless: a spouse or partner, his guardian, his children, his friends.

I ask him where he slept the night before. If it was an emergency shelter, I ask him to name it. If it was a transitional housing facility, I ask him to name it. He's been staying in a hotel or motel, couch surfing, or sleeping in a place not meant for human habitation, like a vehicle or the woods. He's been in jail, in a mental health or substance abuse facility, or the hospital.

I ask him how long he's been homeless this time. I ask him how many times he's lived on the streets or in emergency shelters in the last three years, including at present. Whether, when he combines all the times he's been homeless in the last three years, it equals twelve months or more.

I ask him the primary reason he became homeless. He has lost his job. Or has alcohol or drug problems. He's turned twenty-one and is forced to leave foster care. He's left to escape abuse. Or is a recent immigrant. He has lost his home through foreclosure. He has mental health or emotional problems. Medical problems. He has a criminal history. He's been evicted. He is running away from foster care. There was a family break-up. A disaster, natural or otherwise. He doesn't know. He refuses to answer.

I give him the bus pass. I thank him for his time.

We get back in the sedan and drive to the Mirror Lake Library. We park under a banyan tree and run through the rain up the front steps to the third-floor computer room. I approach each person sitting at a computer and ask in a whisper if he or she is currently experiencing homelessness. I recognize two people from the Saturday breakfasts. I meet a man who didn't sleep last night because he spent the whole night walking around the city—he's been homeless for a day. I meet a dreadlocked kid who quit his "corporate" job in Rhode Island to busk on the streets of St. Petersburg. I kneel next to a woman who yells at me to leave her alone and continues yelling even as I move on to the next person. I meet a woman who is living in a domestic violence shelter with four children. Downstairs, I see a woman I know from the breakfasts. Her face is red and she's missing a front tooth. Her eyes are perpetually half closed, and she's always half smiling to herself, a smile that disappears if she sees you looking.

In addition to numbers gathered during the street survey, the Homeless Leadership Board gathers numbers from emergency shelters, the school board, and the Pinellas County jail. However, not all those numbers are reported to HUD. In 2015, there were an additional 388 individuals in the street survey, 408 in the jail data, and 2,670 in the school data that did not meet HUD criteria for homelessness. Whereas the Homeless Leadership Board counted 6,853 homeless individuals that year, only 3,387 were reported to HUD. Whereas according to HUD, 35,900 people were homeless in the state of Florida in 2015, a statewide count by the Florida Department of Children and Families of school data the year before identified 71,446 homeless children alone.

The difference lies in definition. While HUD includes in its count children and families with unstable housing situations, it qualifies its "homeless" criteria with the requirement that individuals have not been on a lease and had to move more than once in the past sixty days, and will continue to be unhoused due to disabilities or barriers to employment. This excludes anyone who, for instance, lost his or her apartment last week and is living with family; individuals whose barrier to employment is not a disability but their lack of a diploma or their race; those who are employed, but underpaid; and those who've been lucky to move only once in two months but are still housing unstable.

The Florida Department of Education uses a slightly different definition. Its has no numeric or temporal definitions, or requirements that those counted validate the reasons behind their unstable housing. They only ask if those counted are either sharing housing with others; living in hotels, motels, RV parks, or campgrounds; living in shelters; have been abandoned in a hospital or are awaiting foster care; are living somewhere where human beings shouldn't live, like cars, parks, bus stations, or abandoned houses; or are migrants who haven't yet settled in a new place.

Family homelessness has been on the rise as low wages fall further still while housing costs skyrocket. In February 2016, President Obama proposed putting $11 billion in federal funds into fighting family homelessness over the next ten years—up from $4.5 billion in 2015. More than half of homeless families with children live in five states—New York, California, Massachusetts, Florida, and Texas. But compared to the Florida Department of Children and Family's 2015 finding that over 77,000 of Florida's schoolchildren were homeless, HUD reported that there were only about 123,000 homeless children nationwide. Data collected by HUD helped shape Obama's $11 billion proposal, including the Family Options Study, an analysis that compares the effectiveness of housing vouchers, or long-term help, to "rapid rehousing," or short-term assistance. It found not only that vouchers were more effective, but also that the cost to communities was the same.

The inciting incident in G.W.'s novel is the death of Branna Coyle. Branna is a twenty-one-year-old woman with mental retardation who is prone to seizures. She's the daughter of Molly Coyle, who, according to the G.W. character, calls him "for just about everything." Molly Coyle and her eight children live on the streets. Branna is the youngest and breaks everything she can—especially things that shine. Ashtrays. Glass tables. She and Molly were evicted from their last apartment because Branna broke the front window. The landlord called them animals.

The conflict of "Things That Break" is its setting, and the character of G.W. is the lens through whom we view it. When the fictional G.W. gets off the street, he appoints himself the mayor of Williams Park. There, he mediates between the park's homeless and the city, the homeless and the fellow homeless, the homeless and God. He grows close to the Coyles because they need him. When Branna dies, he's there.

"They are young people on the verge of going under," he says in the novel of the Coyle children. Molly's daughter Collete goes to her mother's spot to wake her on a Sunday. She's brought coffee. Her mother wakes, but Branna doesn't. Collete calls G.W. "I can't really do anything for them but be there," he says. He takes a bus to Williams Park. He thinks about how they called for Christ after Lazarus died. He agrees to hold a service for Branna that evening.

The service is held in the park. All the park's homeless are in attendance. They pass around a bottle of whiskey. It's been a hard day for everyone, unraveling secrets and testing the limits of each character's morality. G.W. speaks to the people who are gathered there. "I want you to know that the word 'repent' means to 'turn,'" he says. "Change course. Don't do what you used to do."

He places a megaphone to his lips.

"There are some who do not have the power to willfully 'turn,' to change, to be willing to repent . . . Ask yourselves, all of you, if given the choice to turn, to change, if our sister would not have been the first in line to have an average childhood and live an average life."

He slugs the whiskey.

He holds a vigil for all who have died on the streets before Branna. "The souls of our comrades, plus the souls of our kin, have gone before us to loosen our bonds and show us the way," he says.

He gives Communion. Each person comes up reluctantly and takes the bread and wine. Then they disappear into the park, and G.W. finds himself alone. He thinks about what he's just done. He is satisfied and grateful that he's done Branna right.

"Things That Break" is a book about family, about the ties that bind people to people: mother to child, pastor to disciple, lover to lover, friend to friend. It queries the ties that bind character to reader—for the cruel world the characters inhabit is one that implicates the reader in its structure. And yet it asks us to empathize. It tells us to repent. It instructs us to humble ourselves and care for one another.

Two days after the homeless count, G.W. calls to tell me he's relapsed again. Less than a month has passed since his heart attack. He's spent sixty dollars that wasn't his to spend. He asks if I can replace it. I tell him no.

Two days pass. We don't speak. I worry that he's using. I text him and he doesn't answer. He always answers texts. I worry that his silence means that he's dead. I think about the people in his house finding him, how many days could pass before they think to go in his room. What that would mean for them. What it would mean for Missio Dei. I worry about his book. I make a silent promise to him that it will see the light of day.

He answers the phone at nine a.m. on the morning of the third day. He's slept through the last three days. Someone paid back the money. He regrets what he's done. He's tired. I apologize for reaming him out, and tell him I was scared. He says he's scared, too.

That Saturday, I arrive late to the church. G.W. is in the kitchen with a troupe of co-ed scouts in matching T-shirts who are helping cook. It's near the end of the month and government checks have run out, and two hundred people are lined up outside, waiting to eat. When the food is ready, we wheel it out to the tables and make a circle. G.W. speaks.

"The face of this breakfast might change," he says. "I managed, last night, to get a donor to prep the coffee and stuff like that. But otherwise, we're living off the land. It seems as though there are insurmountable odds against us having breakfast."

G.W. tells us that Trinity Lutheran has raised Missio Dei's rent. They also want to have the food brought in from now on, instead of being prepped in the kitchen. That just won't work, G.W. says. "What would happen to all of you? What would happen to my volunteers?"

He recounts that something similar happened at the Unitarian church years prior. They served dinner there. The church members decided that they wanted to cook and serve instead of the homeless themselves. "They wanted to work and pat people on the head, and say, 'Bless their little hearts,'" G.W. says. "This is the antithesis of what I'm about. I'm about enabling people to bring themselves up from the shit. This is just a way station," he says. "This is not our reality."

From now on, whenever someone passes a supermarket, they are to stop inside and ask for extra food. Dated cereal. Dated milk. Plates and napkins. Tell the people inside about the breakfast. Lean on the community.

"There are great odds against us, even in this building," G.W.

says. "But I'm sixty years old and I seldom lose. A man who was seventy-three years old, when most lives are over, got released from prison and changed the world, and I'm talking about Nelson Mandela. They thought he was drained. They thought he had shot his last shot. But he is one of the most famous men in the world for what he did—for other people. So I'm not going to aspire to do anything less."

We hold hands.

"Lord, thank you," he says. "Thank you for today. Thank you for our strengths. And thank you, thank you, thank you for our weaknesses, because they show us that we come and we go, and we can progress. Because where we are now, we're not going to always be. And thank you, God, that we're not where we used to be. Lord, help us. We see your hand in this. We see you all the way. Lord, stand with us. Help us, and let us kick some ass. In Jesus's name, I pray these things. Amen."

"Amen," we say.

A Tiny Homes Project volunteer meet-up is held that afternoon at a downtown pastry shop. Fifty people crowd inside—more than a hundred signed up after reading about the project on the cover of the *Tampa Tribune*—and eat crème brûlée and key lime pie, telling one another why they're there. A real estate agent who's been bringing sandwiches to homeless people in Clearwater for years says she hopes to help Celebrate Outreach work with the city on zoning regulations. A man who spent three years on the streets of Orlando after the corporate job he'd held for two decades was downsized says he wants to do whatever Celebrate Outreach needs him to do. Another man just wants to learn to build houses.

George Bolden wears a cream-colored suit as he smiles and shakes hands. A week ago, he held a conference call with students in the architecture program at the University of South Florida who are helping design the Tiny Homes with input from veterans. At the end of the semester, students will submit their designs and the future Tiny Homes residents will vote on the ones they like best. Each will be under five hundred square feet, use renewable energy, and cost less than $25,000 to build. Veterans will pay around $250 a month toward ownership—with financial help and supportive services available if they need them. Celebrate Outreach has identified some city-owned lots, each between $1,000 and $3,000. They hope to buy at least one the following week. They'll start building in September.

"There was a question asked on our Facebook page about what role children can play," Bolden says, settling the room. "Perhaps some of our volunteers today can come up with specific suggestions for things children and youth can do."

"Water boys," one woman offers.

"Helping in the garden," says another.

"I'm just thinking about my own kids," says a voice behind me. Everyone turns. A woman with wet blue eyes leans against the door frame with her arms crossed over her chest. She wears a woven white sweater and faded jeans. Her hair tumbles down over her shoulders. Her mouth trembles. "They love learning how to do things, like . . . They would love to do the tile. You know, a veteran, if it's not perfect, they're not going to care. They'd rather see a fingerprint in the grout. It's homemade."

Bolden thanks her. Everyone turns back around. Someone else suggests painting. Another, babysitting.

I find the woman later, sitting alone on a leather sofa. Her name is Megan, and she is the first recipient of a Celebrate Outreach Tiny Home. Megan is a fifty-year-old mother of three and a recent divorcée. She suffers from PTSD, military sexual trauma

(MST), and bipolar disorder. She is two and a half years sober and, after leaving the air force, worked as a nurse in a neonatal intensive care unit for twenty-five years—until September 2015, when she Baker Acted herself. The Florida Mental Health Act of 1971, or Baker Act, allows for the involuntary examination and institutionalization of an individual with mental illness who may present a danger to themselves or others. Megan was suicidal. "I did the best I could, and things started falling apart anyways," she tells me. Her car was repossessed. She lost her family. She was about to lose her home. "But somewhere in all this chaos, I asked for help. I asked a higher power for a will to live."

She went into PTSD and MST treatment at the VA. She found people there who understood her pain, who didn't question her triggers or ask her why her life was falling apart in spite of her best efforts. When a friend from the group heard she was losing her house, she put her in touch with Celebrate Outreach.

"They reached out to me two days ago." Megan laughs. "I have no idea what to expect." But listening to George Bolden talk, she began to fantasize about drawing up a plan for her new house. "I'm going to learn how to do carpentry," she says. "I'm going to learn how to put tile together. I'm going to do all the things I've always wanted to do but couldn't because they weren't girly."

She'd been living on Family and Medical Leave Act payments and then on short-term disability insurance since Baker Acting herself in September. She grieves the loss of her job but admits that she was burned out. Neonatal intensive care is triggering for her PTSD: she might have somebody die in her hands and somebody else born in her hands in the same day. The bureaucracy of the health care system is also triggering. She worked nights, weekends, and overtime, constantly. She witnessed decisions being made that she felt were unethical. "We don't really have ethics committees anymore, because you can't have them and try to keep babies alive that probably should have been let go," she says.

The hospital was short-staffed. On the outside, it might look fine, but it's "rotten as hell" on the inside. "Hospitals want their employees to lie and make everything look pretty," she says. "When you lie to yourself, that's self-abuse. When you lie for somebody else, it's self-abuse. We're enabling the medical system to hold our health hostage. I get angry over that. Everybody's getting set up to lose."

It reminds her of what she went through at Chanute, the air force base where she was stationed. "You have people with political power, with money, making decisions," she says, of the hospital. "But it won't be the people that are making decisions that are held accountable, it'll be the little people. And they're trying to protect themselves."

She's been living with the pain of Chanute for thirty years. Her first trauma happened in 1987, others in 1990. She never allowed herself to think about it after she left. She could never afford to. She supported her family financially, her husband and three daughters. "I thought, in order to live with the trauma, 'Well, I'm just going to work really hard and be really good,'" she says. "That's what I did, and I was. Until I wasn't."

Of all that she's lost, she grieves the most for her family. Her daughters live with their father now. Two are in their teens. The youngest one struggles with her mother's situation the most. "I think it really hurts her because I isolate [myself] from them," Megan says. She doesn't know if she does so because she's in pain or ashamed. "I just have to get brave and face it. I try to do a little bit at a time. I just try to have faith that there will be a couple of good memories, that it won't be all lost."

The experience of losing her car was humbling, though she admits it was also liberating—it gave her a sense of peace. There are no more car insurance payments, no more gas. She gets around now on a bicycle and a bus pass. She's learning to be content with what she has, to live with less. She's also learning how to be selfish. "Me asking for help helps somebody else, and helping somebody else helps me," she says. "Seeing things through other people's eyes makes you connect with them, be empathetic to them. They become human. They become real. That's when you start seeing value."

Help for veterans like Megan through the Tiny Homes Project includes support group meetings, sessions with psychologists, and service dogs for veterans who want them. The dogs are trained to help treat PTSD. There's also a gardening element to the project.

"Gardening provides food, so you're ending hunger," George Bolden tells me later, after the café has cleared. "Gardening also creates entrepreneurship opportunities. You can sell the produce to local restaurants, institutions, hospitals, schools, or neighbors, other families in the area." Gardening is also therapeutic. Once you've gone through the trauma of taking life in a combat situation, he says, to nurture a plant and watch it grow into a head of lettuce, or a tomato, or a cucumber is incredible. It restores a person's humanity, gives them the joy of caring for life. His hope is for each tiny home to have a garden of its own and for there to be a community garden in each village.

A few days later, as I'm leaving town, G.W. texts to say he has something to tell me. He won't tell me over the phone—he insists I inform him when I'm back home in New York so he can text it to me from Florida. I text him as I walk through the door of my apartment. He replies, "I think I over-exerted myself last Sunday," referring to the last time I saw him, at Sunday dinner. We'd finished serving, and he had emerged from the fellowship hall elevator and shuffled across the tile floor, his breathing shallow. His face had been shiny, his gaze distant. He'd sat down heavily in a chair and nodded to me. He'd told me he wanted to start doing yoga.

"I'd been on my ass literally all week," he texts me now. "And then I, full of blithe stupidity, drank three beers on Friday. I had another attack, a small one. But I just can't do it anymore. I'm trying to get my head on top of the hill. Hang with me."

He hasn't taken himself to the hospital because he doesn't want to go. He's been sleeping for days. He says he just wants to rest. It occurs to me that rest might be what he's wanted all along.

With the help of Celebrate Outreach's connections, the meals have been relocated from Trinity Lutheran to Christ Methodist, down the street. This past Sunday's was the last to be held at Trinity Lutheran. For the time being, the Missio Dei church is homeless. But they'll find another place to live, and more money. This has happened before, and he's used to long odds.

Two weeks later, G.W. sends me another text: "Hey! I'm writing. How about you?"

Sunshine State

Characters in the following story are presenting their own versions of events and do not necessarily reflect the truth, which we may never know. Some names have been changed to protect anonymity.

The Suncoast Seabird Sanctuary sits hidden behind a bank of palms on a curve in Gulf Boulevard, which follows a strip of barrier islands from the top to the bottom of Pinellas County, Florida. The curve marks the place where Redington Shores stops and Indian Shores begins. Pastel-colored luxury hotels and salty stilt houses flank the sanctuary, and a giant fiberglass pelican marks its entrance, its paint chipped and faded from the sun. Next to the pelican a rainbow flag reads "OPEN."

Growing up in Pinellas County, I visited the sanctuary many times as a child. I remembered it being a place of encounters—with strange species, with wild instincts. Standing in the faintly shit-scented gift shop on my first day as a volunteer, I told the coordinator, Adrianne Beitl, that I'd returned to write about it. I'd be working at the sanctuary for six weeks, doing research.

"Are you a journalist?" she asked.

"I'm more of a memoirist."

Adrianne had hair the color of wet sand and a face that looked perpetually confused. I filled out an application and handed it to her. We looked at each other.

"There have been bad things written about us in the paper," she said.

I thought I would write an essay about birds. In a lightly magical way, I'd begun to notice them all around me. A few weeks before, I'd found a fledgling pigeon on the sidewalk in my Brooklyn neighborhood. Its feathers were coming in and it couldn't yet fly, and I'd feared for its safety on the busy sidewalk, unable to find its nest. I brought it to a bird hospital on the Upper East Side. In the examining room, a veterinarian fished trichomoniasis buildup out of its throat with a cotton swab on an orangewood stick and, in the custom of the shelter, named the fledgling Sarah. I felt I was on the trail of something ancient. I felt I was hunting something powerful and primal.

The Suncoast Seabird Sanctuary was founded in 1971 when Ralph Heath, a newly graduated premed zoology student on his way to do some Christmas shopping, brought home a cormorant he'd found dragging a broken wing on the side of Gulf Boulevard. At the time, Ralph was living with his wife, Linda, and his parents in his robin's-egg-colored childhood home, which later became, and remains, the sanctuary office. With the help of a local veterinarian, Ralph and his father nursed the bird back to health, named it Maynard, and let it live in their front yard. Soon after, a local bait fisherman caught wind of their success and brought them an injured gull. Then the postman left a bird on their doorstep. Then birds began arriving outside their front door every day.

Ralph recounts the story in *Flying Free 2*, a low-budget video history of the sanctuary—an experiment in nontraditional camera angles, somehow costarring local newscaster Bill Murphy. Like old friends, clad in matching safari shirts, the two sit across from each other on wood pilings inside the brown pelican enclosure and engage in what sounds like a casually overrehearsed conversation about Ralph's history.

"Ralph, let's talk a bit about someone who had such an impression on you and the direction of your life," says Bill, looking straight into the camera. "Talk a little bit about your dad."

"My father was a very unusual doctor," Ralph replies, as if hearing the question for the first time. "He was an MD, not a veterinarian. But he was probably the best surgeon Tampa ever had. He could put anything with tissue back together."

Now in his seventies, Ralph has lived on sanctuary property for more than fifty years. His house is across the street from the sanctuary office and overlooks the Gulf of Mexico—he lives on the bottom floor. The first time I saw him, Ralph was wandering down the dirt pathway in front of the hospital, shirtless and in flip-flops, speaking into a flip phone, saying, "We haven't really had money to do that." I'd assumed he'd been removed as the sanctuary's director, having followed up on Adrianne's recommended readings: Ralph had been in the news a lot over the last decade, for all the wrong reasons. So I asked another volunteer, an older, short-haired woman, who this slovenly, beer-gutted man was, as he seemed to be making decisions. Just moments before, she'd offered to dissect an owl pellet for me, so I'd decided she was friendly.

"That's the founder of the sanctuary, Ralph Heath," she said. "He's a very strange man."

"So, he's here a lot?" I asked.

"More than we want him to be."

I watched Ralph shuffle toward the office on his phone. His hair was thin and uncombed, and patches of scabs and scars dotted his bare skin. He paused in front of the vacant visitor information counter to look at a night heron perched on the roof like a gargoyle. The bird stared back down at him.

"What does he do?" I asked.

"Nothing. He drinks a lot, if I had to guess. So, if you see Ralph Heath, founder of the sanctuary, in the newspaper," she concluded, "you'll know to stay away."

The cover of the August 1974 issue of *Smithsonian* shows a blue heron standing on a grassy bank in front of a calm lake. A hunter's arrow dangles from its throat. Below it, the caption directs readers to an article on page thirty, titled "Volunteers Rescue Injured Wildfowl." The article is richly photographed. In one picture, Linda Heath's delicate hands slip a pill inside a baitfish to serve to a sick cormorant. In another, a hawk with a bandaged wing perches atop its cage, as if in thought. The article says Ralph and Linda "have become experts in the ways, deliberate or inadvertent, that birds come to grief at the hands of man." Linda cares for the baby birds in the hospital while Ralph fashions prosthetic limbs for birds whose legs must be amputated. "Heath is at his best saving birds which his fellow citizens have damaged," the writer asserts.

That year, the Suncoast Seabird Sanctuary became the first facility in history to mate the brown pelican in captivity. After years of exposure to pesticides and pollutants, the bird was endangered and on the verge of extinction. By the following year, the sanctuary had hatched the first brown pelican egg and Dewar's Scotch had featured Ralph in their Dewar's Profile ad campaign. According to the ad, which ran nationwide in publications like *Esquire* and *Playboy*,

Ralph's hobbies included restoring antique cars and filmmaking. His favorite book: Rachel Carson's *Silent Spring*. In the ad's photo, he lifts a healthy pelican onto his forearm, its wings raised in preparation for flight before a virgin shoreline. Ralph is young and muscular, his dark hair full and perfectly mussed, '70s moustache responsibly trimmed yet wild. His gaze meets the camera as if he's about to speak; his hand ever so gently touches the pelican's soft breast. The two make a perfect couple, an iconic union: man and bird.

20/20 did a special on Ralph and the sanctuary, then the *Today* show and the *New York Times*. Editorials showed up in the local papers every few days with headlines like "Birds: Our Responsibility" and quotes from Ralph about the value of animal life and the public duty of seabird ministration. Disney came to the sanctuary seeking animal stars for its Discovery Island theme park. Soon, Ralph was sending rehabilitated birds to zoos all over the world, in Greece and Singapore, Spain and Barbados. On a trip to deliver pelicans to Texas, Navy Commander and president of the local Audubon Society Bruce McCandless, the first man to float free from the *Challenger* shuttle—"all atilt, with no tether," as Ralph later put it—took him bird-watching in his amphibian aircraft. Ralph's reputation as a wildlife documentarian grew as he shot films from the sanctuary yacht, the *Whisker*, and fully inhabited his role as an advocate for the plight of birds everywhere.

Ralph and Linda divorced three years after the sanctuary opened, and in 1982 he married Beatrice Busch, millionaire Anheuser-Busch heiress, wildlife photojournalist, and world traveler.

Beatrice had spent winters in Tampa Bay with her beer magnate father, August A. "Gussie" Busch, also president of the St. Louis Cardinals. While at the family's beach house in Pass-a-Grille she came across an article about the sanctuary in one of the local papers and recalled hearing the captain of one of their boats talk about Ralph. Beatrice had worked for a time at an animal orphanage in Nairobi and had just finished filming a television special about the migration of humpback whales with her filmmaker ex-husband. She was interested in people who were interested in animals, and decided to pay Ralph a visit.

She arrived just as Ralph was leaving to rescue a pelican hooked by a fisherman, and on a whim accompanied him. Despite the dirty work involved—they were out in the muddy mangroves in tennis shoes, grabbing birds with their hands, getting nipped by cormorants—it was love at first flight for Beatrice and Ralph.

Their wedding took place a year later on Grant's Farm, a three-hundred-acre St. Louis estate that was once the home of Ulysses S. Grant and on which Gussie had raised seven of his ten children. On the farm sits a cabin hand built by President Grant, a thirty-four-room neo-Renaissance château, the terrace of which was the site of the ceremony, and a fully operational private zoo, which included chimpanzees and elephants Gussie had trained himself. Within three years of marrying Ralph, Beatrice had given birth to three sons: Andrew, the oldest, and twins Alexander and Peter. Within another two years, Ralph and Beatrice had separated. She'd been rehabilitating birds and traveling with Ralph to promote the sanctuary as well as spending ample amounts of time with the animals at Busch Gardens. But she was ready to move back to St. Louis and be with family. Ralph refused to go with her. He couldn't leave his birds.

In the beginning, the birds had united Ralph and Beatrice. After the wedding at Grant's Farm, they could have left on honeymoon to Switzerland—the rest of her family resided there and Beatrice wanted Ralph to meet them. But instead, they postponed their trip until the St. Petersburg City Council reached a verdict on whether to let Ralph expand the sanctuary onto a 125-acre plot of land in the environmentally sensitive Gateway area, called Sod Farm. It was a plan he'd been working on since 1974, when he'd gotten into the habit of buying land as a way to preserve it against booming development along Florida's coasts, leaving it wild for the birds. Now Ralph wanted to build a zoological park on Sod Farm, and had community support. But the city wanted to zone the land for industrial use. They were expected to come to a decision in December.

The story had monopolized Tampa Bay news circuits for the last several months, and petitions in favor of the park collected hundreds of names. Ralph and City Manager Alan Harvey faced off in the headlines: Ralph claimed he'd been "double-crossed," as Harvey first had promised him the plot and then retracted his offer when he realized just how much land the city stood to lose. Reading an impassioned statement at a city council meeting that fall, Heath urged the public to stand against any government employees who didn't behave as environmental stewards, claiming such people couldn't possibly hold their constituency's best interests in mind. "These individuals do not yet realize that mankind and wildlife live in the same ecosystem," Ralph declared. "If we destroy the environment for the wildlife, we destroy the human race as well." In the end, the city turned the plot into a landfill.

People came from everywhere to see the birds. At its height, over a hundred thousand individuals and ten thousand birds entered the sanctuary's grounds every year, making it the largest nonprofit wild bird hospital in the country. It hosted events and presentations every week by organizations like the Florida Native Plant Society and Eckerd College; gave regularly scheduled tours; visited schools; published a quarterly newsletter, *Fly Free*; and held wild bird photography workshops. At its largest the sanctuary employed forty-four workers, including an internationally praised bird rehabilitator, Barbara Suto; a staff veterinarian working in state-of-the-art emergency facilities and a surgical center; licensed veterinary assistants; groundskeepers; office staff; a marketing director; an all-volunteer rescue team bringing in up to forty-five new birds every day in pickups emblazoned with the sanctuary logo; and volunteers keeping tight shifts on the grounds and in the sanctuary hospital. Ralph was the face of it all, the bird god of Indian Shores. It seemed as though nothing could bring him down.

When Chris Walls, the sanctuary's groundskeeper, showed up for work on the morning of March 4, 2013, he was met with an unusual sight. On a normal day, the sanctuary hospital would have already been open and Barbara Suto, the hospital director, would be checking on her patients. Matthew McDermott would be making rounds of the enclosures, doling out fish to the pelicans and shore birds, mealworms and seed to the perchers, rats to the owls, and raw chicken to the raptors. Micki Eslick would be unlocking the sanctuary office and going about her morning paperwork routine with her son, Donnie. There would be volunteers at the outdoor laundry folding towels to deliver to the hospital. The place would be bustling. But that day, the sanctuary was a virtual ghost town. In the hospital sat twenty-four sets of keys, each bearing a Post-it resignation. Along with his boss, avian supervisor Greg Vaughan, and Greg's wife, Kellie, Chris was one of only three people there.

I wanted to talk to Chris about that day. I found him leaving the wading bird enclosure on his way to the outdoor hospital with an empty fish bucket. Behind him, a sandhill crane stalked along the far edge of its concrete water pool like a ghostly ballerina. Nearby, other wading birds called from their shady corners: blue herons, egrets, gulls, white ibises. Chris wore knee-high rubber boots and fishing waders. He has a patchy beard and scruffy dirty-brown hair. He's tall and gregarious, but humble, glad simply to have an occupation. We talked as he knelt in the dirt over a stack of feeding charts in the outdoor hospital, an area with two rows of cages separated from the rest of the sanctuary by a wooden fence, where seabirds are sent to rest further before release. A few feet away, a blue heron was nursing a broken wing, and three enclosures housed juvenile night herons and snowy egrets separately.

There were now six paid employees working at the sanctuary. Chris had begun four years ago, doing community service for a Roxycontin charge. He banged out fifty-five hours of community service in two and a half weeks, doing laundry and grounds repairs, then stayed on as a volunteer as he was looking for paid work. When another employee quit, Chris was given his job. Even in 2012, he said, the situation was far from perfect. A lot of employees weren't getting paid, and some of them were falling behind on their rent. People were steadily leaving jobs at the sanctuary.

Around this time Ralph was accused of stealing money out of the donation boxes—outed to the press by former employees. The following year, Ralph announced the sanctuary would no longer be rescuing birds and would be transitioning to a volunteer-based staff due to its inability to keep up with debts. The IRS was pursuing them for almost $200,000 in unpaid payroll taxes. They owed more than $21,000 in employee back pay. All of it made the papers. Then, in November 2013, Ralph was arrested and charged with workers' compensation fraud for failing to pay workers' comp insurance for seven months the previous year. During that time, a man had slipped in a truck while delivering ice at the sanctuary and had gone to the hospital.

I asked Chris why people weren't getting paid.

"It was the management," he said. "Our CFO at the time didn't know what she was doing. She always got paid, her son always got paid, but no one else got paid."

This was Micki Eslick, who denied any wrongdoing when I later interviewed her. I'd seen her in *Flying Free 2*. She appears in a bizarre interlude, though she's not identified as the CFO of the sanctuary but rather as Ralph's cousin. She wears French-tipped acrylics and has short black hair. In a smoker's rasp, she tells a story about ten-year-old Ralph.

"If you can remember, back in the day," she says, as if to me personally, "the chicken farmers would put wooden eggs in the chicken pens so the chicken snakes would go in and eat the wooden eggs. Then they'd go off and die, and not bother the chickens."

A patient of Ralph's father brought in a snake with a wooden egg in its belly, and Ralph knew it would die unless they operated on it, which they did, then kept it while it recuperated. All of their lives, Micki says, every time she can remember, Ralph has been saving animals.

"Anything wildlife," she clarifies, "he's always wanted to help." End segment.

Where is Micki now? Chris ignored my question.

"Now it's Kellie," he said. "She's managing the money."

The sanctuary currently relies heavily on outside help. Licensed rehabbers pick birds up from the hospital a few times a week so they can receive care the sanctuary can't provide, as it no longer has a veterinarian or any licensed rehabbers on staff, aside from Ralph. Each rehabber specializes in a particular species. The Audubon Society of Clearwater helps. There's a duck lady, Mary. Then there's Eddie Gayton, who exercises the sanctuary's hawks and takes their screech owls. It seemed obvious now, but it hadn't occurred to me until then that the woman running the hospital, Shelley Vickery, wasn't a veterinarian. She's a former kindergarten teacher.

A rotating cast of volunteers bolsters the work of the small team of paid employees now carrying out the sanctuary's daily operations. Some have been there for years. Others work there for a few months while they find some life direction. I spent most of my time in the indoor hospital scrubbing cages full of fish-scented shit and squirting vitamin-and-mealworm mash into the mouths of baby birds with a plastic syringe. I was told on my first day that feeding baby birds was easy—I just had to aim at the side of the throat, away from the windpipe. This wasn't always the case: sanctuary volunteers used to go through special training to feed baby birds, in part due to how easily they imprint on humans. They grow accustomed to seeing human faces every day, and hearing their voices, and being cared for. They will never learn to take care of themselves.

I was feeding a pair of baby blue jays when Greg came into the hospital chuckling. Shelley had just called after releasing a baby Muscovy duck back into the wild and had sworn off releasing birds ever again. "She put him down on the bank, watched him swim out," Greg said. "Otter come right up from underneath and ate him!" He laughed with Kim, a veteran volunteer, at the thought of poor Shelley.

"Is that mourning dove here?" he asked, calming down. "I hid it last night so Ralph wouldn't get to it."

The dove was still there, Kim said. Greg was glad.

"Why would you have to hide it?" I asked.

"So Ralph doesn't take it home with him," said Greg.

"He's an OCD. Hoarder," Kim explained. "We don't talk about it."

I continued about my business as they finished their conversation, struck by Kim's casual tone. Hoarding animals was nothing to laugh about. I weighed the open secret of Ralph's alleged hoarding against my knowledge of the sanctuary's finances. Beyond their inability to maintain a full-time, permanent staff, in 2011 Ralph signed over the deeds for the sanctuary property and his house to Seaside Land Investments LLC, a Texas-based corporation owned by his sons—heirs to the Anheuser-Busch fortune and the fortune of their stepfather, Adalbert "Adie" von Gontard III.

I knew that all three of Ralph's sons were in town. I'd also heard that Ralph's best friend, Jimbo Guastella, had recently moved into the second floor of his house. Before leaving for the day, I asked Kim if she'd ever seen inside it.

"I caught a glimpse of it once," she said. "It's wall-to-wall pigeons."

Pigeons are an invasive species in Florida. If you find an injured pigeon, you can take it to a licensed rehabber, who can rehabilitate it but can't legally release it back into the wild. Some rehabbers will choose to euthanize. Others rehabilitate and rehome them. With nice facilities some rehabbers can accommodate as many as fifteen or twenty pigeons at once.

So Ralph was taking the sanctuary's pigeons. All of them, apparently.

"But the mourning dove?" I asked. "They're a Florida native species."

Kim shrugged. "Well, yeah. He comes in and looks around."

THE COSTS ASSOCIATED with daily rescuing, feeding, and housing of four hundred birds are considerable. There's the daily cost of six hundred pounds of fish and their refrigeration in two walk-in units. Then there's medicine and staff. Facilities and repairs. Rescue vehicles, both land and water. In times of emergency, such as oil spills, red tides, and hurricanes, the costs skyrocket. Not to mention the labor demands. Greg Vaughan was working at the sanctuary when the

Deepwater Horizon oil spill happened. The sanctuary alone received 120 gannets, large shorebirds not used to walking on land. The birds had lost their feathers after countless Dawn washings stripped them of their delicate oils, so the sanctuary kept the birds for a year while the feathers grew back. Walking on land can give gannets blisters, so each had to wear special shoes: mesh flip-flops taped to their feet. Then they were placed in water for four hours a day in a specially built enclosure that kept them from climbing out. It doubled the workload, Greg said, and nearly doubled the cost of food.

The sanctuary has never applied for government grants, claiming there's too much paperwork involved. It has opted instead to rely on private donations from local philanthropists and bird lovers. Some come in single-bill kindnesses dropped into the sanctuary's pelican-shaped donation boxes; some are quite large, whether monetary or in the form of cars and yachts. But there have always been low periods. At first, the *St. Petersburg Times* (now the *Tampa Bay Times*) and the now-defunct *Evening Independent* were supportive through periods of scarcity, urging local citizens to pitch in where they could. When Dick Bothwell of the *Evening Independent* reported on the 1974 *Smithsonian* article, for instance, he added accusingly, "If everybody who loves birds in the Tampa Bay area would put their money where their bills are, the sanctuary's troubles would be over, for a time at least."

And yet there were years of unimaginable prosperity. Every few months, someone died and left millions of dollars to the sanctuary. Others were all too ready to donate when asked. Ralph's mother, Helen, managed the business and kept up positive relations with the public, faithfully sending thank-you cards and developing the sanctuary's Adopt-a-Bird program, managing memberships and keeping precise books until she fell ill and Micki took over. Ralph met his best friend, Jimbo, when Jimbo's band held a benefit concert for the sanctuary down on St. Pete Beach and raised $28,000 in T-shirt sales alone. Jimbo told me he'd needed to earn some good karma and had been looking for a cause. He and Ralph were fast inseparable.

"I had hair down to here," he reminisced, pointing to his shoulder. "Eighteen different colors from being bleached out, working parasail during the day, and I sang, and there were girls.

"That was part of the bad publicity," he added.

I'd met Jimbo in the sanctuary's gift shop. He's an aging surfer type—leathery but colorful. He was chatting up Diane, a petite middle-aged administrative assistant with blow-dried hair and a permanent low-grade smile. The next evening I observed him singing "Carolina on My Mind" into a beach bar microphone in Gulfport and chatting up a woman in the audience, asking her the name of her young son. I was there because Jimbo had invited me and because I figured Ralph would be there, too, and maybe his sons; I was right

Jimbo had told me that Ralph really liked to spend money. According to him it was part of the sanctuary's downfall. Ralph didn't know how much money the place had because he'd never been interested in the sanctuary's operations; he just happened to be the face of it. With his mother running the office, he never thought to learn how it ran. When publicity went south, so did donations. In 1996, Ralph came under criticism for overstating the effect of a cold winter on shore in a fund-raising letter—he said, wrongfully, that birds faced famine. The next year, his expenditures came into question over his waterfront home, which he'd purchased for $300,000, and the sanctuary yacht, which he'd purchased for $355,000. The yacht had five luxury staterooms, satellite TV, and a hot tub. It was supposed to be a source of revenue for the sanctuary, bringing in charter fees and serving as a venue for donor parties. When it came out in the press that Ralph was taking it down to the Bahamas for months at a time, he claimed he was using it to research the effects of plastic pollution on the ocean floor.

But Jimbo had mentioned the boat to me. He'd told me that Ralph took him down to the Cayman Islands on it, a month at a time, and paid him a hundred dollars a day to play guitar on the beach. Micki went sometimes. They just wanted to party. "I watched the gasoline go in the boat," he said. "I came home with three thousand dollars in my pocket."

Before Jimbo's show, I found Greg and Kellie in the next-door restaurant ordering burgers. They're the kind of people you don't talk about separately; one seems out of place without the other. Maybe it's their thick midwestern accents. Maybe it's Greg's quiet, no-frills attire counterbalanced by Kellie's sundrenched hair and two rings on every finger. They were down in Florida on vacation nine years ago when Greg spotted an ad for a groundskeeper in the paper. He applied and got the job, sold their business back home, and moved down within the month. Mrs. Heath, Ralph's mother, was still running the office, with Micki assisting. When Mrs. Heath fell ill, Kellie, a certified nurse, moved in with her, and Micki took over the office. Kellie was there when Mrs. Heath died. This is how the sanctuary has always been run: not as a business, but as a family operation.

That night at the bar, Greg told me that Ralph owes him and Kellie each $7,000 in unpaid wages—it was an open secret, he said, water under the bridge at this point. Greg had been promoted from avian supervisor to general manager since everybody left. Though Shelley had called Ralph "sort of a consultant" when I'd asked her to explain his current role at the sanctuary, Greg told me Ralph was still the director, despite not doing much of anything. In what sense was he the director? I asked. In the minds of the people? It wasn't clear. What was clear was that his association with the sanctuary had crippled Kellie's ability to maintain donor relations. Too much bad press had hurt them—and it wasn't stopping. Just a few months earlier, Ralph had been in the news again, this time for attempting to sell the same 1963 Corvette Stingray to two different buyers, both of whom were now suing him. He'd collected each of their monies and then refused to turn over the car. These kinds of antics were not unlike Ralph, according to Greg.

"Ralph's an idiot," he said. "You can put that in your book."

He gave me another example: Recently a man had died and left his estate to the sanctuary. Among its holdings were twelve cars, antique and unrestored, but possibly worth some money. Some of the cars were intended for the SPCA of Pinellas County. But when the time came to settle the estate, Ralph said he wanted all of the cars for himself—"for his 'museum,'" Greg said, "which is his warehouse." Ralph fought the SPCA for a year over the cars. Finally, the SPCA backed down and the sanctuary decided to put the cars up for silent auction. Beforehand, Ralph called one of his friends, and asked him to make a bid.

"Ralph's friend made a bid of $80,000 for the twelve cars," Greg said. "And he got beat—a guy bid $90,000."

Ralph threw a fit. He refused to give up the twelve cars. He was taking the man to court.

So now, the lawyer had to be paid, said Greg. "The building that they're in has to have rent on it, the ground has to have rent on it, the car has to have insurance on 'em, the building has to have insurance on 'em."

"The thing is, he's not doing it for the sanctuary," Kellie said.

"Even the attorney said that. Ralph not one time mentioned the sanctuary. It's all, 'I, I, I, I.'"

I looked forward to hearing Ralph's version of events. It had begun to rain lightly by the time we made our way to Jimbo's set in the bar next door, but we stood outside on the concrete step of the open-air entryway and smoked cigarettes, looking at the sunset over the pier, talking casually with glasses of beer at our feet. Adrianne and Diane had come—we waved. Jimbo opened with some classics and got everyone singing along. He wore red Converse high-tops and board shorts, and a red surf-shop tee. He was recovering from a cold and his voice was still raspy, but he sang like it wasn't. When Ralph's son Andrew arrived with his wife and their new baby, Jimbo took a break to talk to them and pinch the baby's cheeks. I stood nearby, listening. Though as yet there was no sign of Ralph, everyone seemed sure he would come at the last minute. He was probably at his warehouse on Starkey Road, they thought. That's where he spent most of his time.

I'd read about the warehouse: there was a foreclosure suit against it back in 2013. The Suncoast Seabird Sanctuary had originally bought the property for $550,000 back in 1997, to be used for storage. Then, in 2009, a local catamaran manufacturer and real estate mogul, Ronald Cooper, issued Ralph a loan for the same amount, with the warehouse as collateral. Ralph and the sanctuary promised to pay $5,500 in monthly interest until November 2011, when they'd make a balloon payment. Cooper received payments on the property until late 2011—then they'd stopped and the rest of the loan had never been repaid. Cooper was patient, but when he found out about the federal tax liens against the Suncoast Seabird Sanctuary properties in 2012, he sued to protect his interests. Then, in December 2013, the warehouse was put up for auction to satisfy an almost $800,000 judgment, which comprised the amount of the original loan as well as unpaid interest, damages, and lawyer fees. At the last minute, the auction was canceled. Ralph's sons registered an LLC named after the Starkey Road address; they listed themselves as officers of the company and paid for the warehouse.

When I'd asked Chris what was stored in the warehouse, he'd told me simply, "You don't want to know." He was wrong. Besides, certain things I already knew: The warehouse measures 35,000 square feet, according to News Channel 8, and has two floors, according to Jimbo. It's owned by the sanctuary but not open to the public. I was determined to find out why.

Ralph arrived at the Gulfport beach bar after Jimbo's set ended, just as everyone was making their way to the bar next door to hear another musician. He wore a dirty white T-shirt with a hole in the sleeve, khaki shorts, and flip-flops, and sat alone on a bar stool facing the stage. Up close, he looked almost childlike—he has an innocent smile and clear, unassuming eyes. The thick head of hair from the Dewar's ad has gone almost bald. He has a big, round belly, and his face is bloated and sunspotted.

I introduced myself as a new volunteer and told him I was writing about the sanctuary. As we talked, I remembered something Greg and Kellie had told me earlier: Ralph refuses to believe what people say about him is true. The things people write in the paper are all lies, he insists; he will prove them all wrong—"The media, as you well know, likes sensationalism," he would explain to me later. Jimbo had echoed Greg and Kellie's sentiment: "He has to win."

Ralph's need to win doesn't manifest as anger. He's actually rather blithe, almost charming—call it stubborn naïveté perhaps. Willful self-deceit. He has a heart of pure gold, and it seems to be true—he's for the birds.

I asked Ralph if he kept any birds at home. He replied that he had one hundred pigeons. Then he told me about the warehouse on Starkey Road, his real joy, where he keeps the other six hundred birds.

Chickens and parrots, pelicans and doves.

Pigeons. Turtles. Furniture.

A few cars.

He's working on a grant to fix the company yacht.

Then he told me something his father had told him when he was a child, a basic principle he'd carried throughout his life: If an animal doesn't have to die, it shouldn't die.

At the end of our conversation, Ralph told me I could call him after the weekend. On the way home, I drove by the warehouse. The gate was closed, but I pulled into the driveway and stood on the pavement, imagining the inside. Were the birds loose and flying around or in cages? Who cleaned this place every day, or every week? Behind the building, live oaks and shrubs crept up to the walls and tangled together. How much money had been poured into building and maintaining the warehouse while the sanctuary couldn't pay its bills?

Here, I thought, standing in the darkness, was the answer to my question: Here is why so many people turned on Ralph.

That was Thursday. On Monday, I awoke to a thunderstorm. Morning storms in Florida are a special kind of sign, a reminder that you're trespassing on Mother Nature's turf—that everything you know could be washed away in an instant. The thunder that morning had made its way into my dreams, drawing me out of sleep. I lay in bed watching lightning flashes across my ceiling, knowing something was wrong. I got up and called Ralph.

"I can't talk now," he said. He sounded frantic. "We got a lot of baby birds in the hospital last night."

It was baby bird season, this was true, and it was also true that baby birds could be blown from their nests by the storm. The hospital tended to get a few more birds than usual on stormy mornings, but certainly not a catastrophic number, as Ralph's tone suggested. The birds were not the issue. He was avoiding me. I asked when I should call him back.

"I'll call you," he said, and hung up on me.

I sipped my coffee.

Greg had given me his phone number the night at the beach bar, telling me we'd talk more when we could be alone. He and Kellie had wanted to divulge things—to expose some goings-on at the sanctuary under cover of anonymity.

I called Greg. No answer. I texted. No reply.

By the time I arrived at the sanctuary that afternoon, the storm had cleared and the heat had risen from the dirt walkways into a thick steam. I found Shelley leaving the indoor hospital and asked about the baby birds. I'd heard they'd gotten a lot that morning.

"Not really," she said, looking confused. I thanked her and continued on toward the office.

Outside, Greg was smoking a cigarette as if nothing was going on—as if he hadn't ignored my call that morning. I bummed a cigarette, acting casual. Took a drag. Pointed across the street at Ralph's house.

"So, that's his?" I asked, as if I didn't already know.

"Yup," he responded.

"Uh-huh."

We smoked.

"Did I miss a call from you this morning?" he asked.

"You did."

"My phone died."

"No problem."

We smoked.

"I won't know anything until Wednesday," he said finally.

I left this mysterious comment hanging in the air between us. We finished our cigarettes and dropped them on the ground, where, looking back, any number of birds could have eaten them. I followed him inside the office like I was allowed to be there and found Jimbo sitting on a couch by the door, holding a glass filled with, I later ascertained, an awful lot of vodka.

Outside, moments later, Jimbo told me everything.

Ralph's sons were dissolving his association with the sanctuary. "They want to keep it alive, and they know it won't stay alive with him involved."

I appreciated this anthropomorphizing of the sanctuary, as if it were a living thing in need of immediate care. Knowing what I did about the sanctuary's troubles over the last few years, most of which could be traced back to Ralph, I wasn't surprised to hear of this development. But I had a feeling there was more to it than money troubles.

"He's so committed to winning," Jimbo continued. "He says it's for the birds, but the birds are suffering."

It wasn't clear whether he was referring to the birds at the sanctuary or the birds in Ralph's house and the warehouse, or even to the birds in general that needed the sanctuary. When Jimbo mentioned that he'd moved in with Ralph to help him clean up his life, I asked him about the birds in Ralph's house. It sat a few hundred feet away from us, and I pictured them inside there now: some caged and some not; some falling in love, as Ralph had described at the beach bar; some sick and waiting to die until he returned home, as he told me on the phone a few days later. From outside, the building looked so small I couldn't imagine how he'd fit a hundred birds inside it. Jimbo answered gravely: "Some don't need to be there. Some are alive only because they're there. And some shouldn't be alive."

Periodically during our conversation, Jimbo would ask me to turn off my recorder so that he could speak candidly, and I would oblige, turning it back on when he signaled it was okay. Deeper into our conversation he decided, for convenience, to leave it on. "I'm just going to trust you," he told me, though I wasn't sure by what standard I should be trusted. I said I knew he wouldn't be telling me any of this if he didn't care.

"I would take—let me be clear about this—I would take a non-fatal bullet for him," he said. "I would really have to think about a fatal one. Because he's old and he's really on the wrong path." He thought about this for a second. "I would jump in front of one only if it wouldn't kill me," he clarified. "I would, because his heart is pure."

I said I believed him.

This was an emotional conversation for Jimbo. He asked me many times while we talked if I was comfortable with him crying, and I assured him I was—that I cry all the time. I'm an artist, like him. He appreciated me opening up to him.

"I'm touched that I'm actually able to be this comfortable," he said. "I'm kind of freaking out. Sometimes I imagine I'm a macho man."

"I knew better from the start," I said.

"He's got birds up there that are so . . . warped," he said, shaking his head. "But they're breathing. And I'm glad about some of them. Some of them should probably be euthanized. But to pull up your flooring so that the birds can shit on the floor without destroying something is a little over the limit, as far as I'm concerned."

I mentioned to Jimbo what Ralph had told me at the bar: that he couldn't let an animal die if he knew it could live.

"I watched him cry for seventy-two hours straight over Snowball dying," Jimbo said. "That was a pigeon. And he didn't shed a tear for his mother."

We looked at each other.

Finally, I ventured: "Do you think his mother's death was just too much for him?"

"No. I think he just doesn't value human life as much as bird life," he said. He sat for a minute, looking at Ralph's house. "He really—He *really* is committed to birds."

Greg walked by, startling Jimbo. "Write that Ralph's an asshole and an idiot!" Greg yelled over his shoulder.

Jimbo was upset. "We've been expressing that differently!" he said. "Wow," he apologized.

"There's some animosity flying around," I said in my most understanding tone. Jimbo agreed. The night before, he'd gone out to dinner with Ralph's sons, to talk about what should be done with the sanctuary. Beatrice had joined via phone. The boys had put their foot down.

"She was on the phone saying, 'We can let Papi have a title. We can call him associate, vice president, director,' which would have made him happy. And Andrew went, 'Fuck no! If he has anything to do with this place, it's going to fail!' And I'm here in the middle, you know? Because I really care about him—Kellie, are you mad at me?" he said as Kellie walked by.

"No!" she said.

"See?" He turned back to me. "I care too much what people think. It's because my parents never cared about me."

That's why he was so protective of Ralph's sons, he explained. Growing up, the boys split their time between Florida and the von Gontard estate, Oxbow Farm, in Virginia. Their stepfather, Adie Senior, adopted them, and they took his name, erasing any public association they had with Ralph. Even back then, Ralph spent most of his time at the warehouse, so the boys had an au pair or went with him to the warehouse when they were in town. But they were lonely and hated the warehouse, Jimbo said. He was around to help them have a normal life.

"I raised those boys," he told me. "They came to me with, 'Please save us from going to the warehouse!' And Andy: 'Take us out! Take us to the beach! Take us to something, do something!' I became their guide."

At Oxbow, Adie Senior made the boys work. They'd get up at five in the morning and work all day, doing manual labor—and they earned money doing it. It taught them work ethic, taught them how to make a dollar, or a few million, the right way. Now in their thirties, the boys have amassed impressive résumés: Andrew is a commercial airline pilot based in St. Louis, has a law degree, and is making his way toward being a judge. In Dallas, Alex and Peter, along with their stepbrother, Adalbert "Adie" von Gontard IV, recently launched an upscale bar called the Eberhard, as well as a dating app called Courtem, which has drawn the interest of several major investors, including venture capitalist Mark Cuban.

"And the twins—this is where the huge fear is going on here right now," said Jimbo. "The twins are in the real estate business of buying failing properties and tearing them apart and making money."

He sniffled. "That's what's making this difficult."

I spent the next week trying to pin down Greg and Kellie. First, Greg would say they could talk before noon. Then, after noon. Then, that evening. Day after day. Was he stringing me along, wasting my time until I had to leave town?

I persisted.

I needed to know what would happen to Ralph and to the sanctuary. Against my better judgment, over three weeks I'd come to care deeply for the place and its people, not to mention the birds.

Finally, Greg called me back, and I picked up a six-pack of alcoholic root beer to bring to the Vaughans' home. They lived down a residential street in an area Floridians fondly describe as "back in the cut," or far from the main road. Greg called to me from the garage as I shut my car door. He sat at a tall, particleboard table smoking a cigarette and attempting to set up his new iPhone. Nearby, a green-cheek conure named Jack nibbled on an apple slice amidst a scattering of birdseed. I asked Greg where I could put the beer and he pointed to a refrigerator, which I found stocked with Budweiser cases. Their friend's liquor store was closing and they'd gotten them for cheap, he explained. Kellie exited the kitchen to join us and I glanced inside—several framed pictures of the Last Supper hung above the sink.

Everyone who talks to me about Ralph is careful to say they admire what he's built. He's done things at the sanctuary over forty-plus years that no one had done before him—fought for the animals when no one else would. He's for the birds one hundred and ten percent, they all say. He's a hero.

"He's just made a lot of stupid mistakes," Greg said.

Kellie nodded. "He just doesn't use his head."

As Greg continued, Kellie stood nearby punctuating his sentences with "Yeps" and "Yeahs," amending or expanding where necessary. They told me that Ralph hadn't been paid by the sanctuary for three years and had been living on social security and the sanctuary's donation boxes—hence the accusations in the press about stealing. His sons were now setting him up to live in the warehouse. It was appointed with beds for people evacuating from the beachfront in case of a hurricane, and Ralph already spent most of his time there. Ralph's sons would pay his bills every month, thus relieving the sanctuary of the responsibility of supporting him and his habits and enabling it to get back on its feet.

I asked what would happen to the birds in Ralph's house.

"He'll have to take care of them," Greg said simply.

"They're going to stay there?"

"They'll go to the warehouse."

"They'll all go to the warehouse," Kellie said.

"Are they okay?"

"They're fine," Greg said.

"They're fine," said Kellie. "Fish and Wildlife check them all the time."

"I mean, yeah, some of them could use more cleaning," said Greg. "But it's not a health issue, anyway."

I was skeptical.

"His house is," Greg said, "but the warehouse isn't. The warehouse is his go-to place. His pride and joy."

It occurred to me that the merging of Ralph's house and the warehouse in his psychological world might have dire consequences. I imagined him left to his own devices with his seven hundred birds in a part of town that was zoned as industrial, isolated from other humans and miles from any supermarket, in precisely the location where his fantasies ran free. The scene had a postapocalyptic wash.

"Isn't there a yacht?" I asked.

Greg laughed. "I'd like to show you that yacht."

"That yacht ain't worth a freakin' nickel," said Kellie.

The *Whisker* was sixty feet long and hadn't been in the water in twenty-one years. It had a wooden hull and was, according to Jimbo, sitting under three feet of bird shit, infested with bees.

"But, uh, I don't know if anybody ever told you about Micki," Greg said.

"Ralph's cousin?"

"Well, she's the one who ruined the place." This was awfully forthcoming for Greg. "It was a power thing for her. She wanted all the power. That's when shit started going downhill."

I'd been curious about Micki, so I was glad she was coming up. As it turned out, shit had been at the bottom of the hill for quite awhile before she entered the scene in 2007. The sanctuary's score on Charity Navigator, a website that rates nonprofits, peaked at one star for years until Micki's predecessor, Michelle Glean Simoneau, took over publicity. The bad press that started the sanctuary's demise had begun long before that, and the sneakiness, too—Ralph's house, for instance, had been purchased by the sanctuary back in 1997, from a funeral director. The plan was to generate revenue for the sanctuary by renting out the house. Then Ralph moved into it.

I tried to imagine, too, what kind of privileges would be afforded to Micki once she had ultimate power over a lowly bird sanctuary—aside, of course, from the two-minute cameo in *Flying Free 2*. But there was history to Micki and Greg, I'd learn. They'd come on at around the same time, before Mrs. Heath fell ill. Micki had retired from Verizon after thirty-four years and come to work at the sanctuary at Mrs. Heath's request. In 2007, Mrs. Heath was ninety-four years old and had been doing the sanctuary's books by hand for almost forty years. Ralph had always been too busy taking care of the birds to learn how to run the business, and he didn't particularly care to learn, either. So his mother had stepped in. And in 2007, she asked Micki to help her with administrative duties.

Micki's husband, Slick, was against the transition—he had worked at the sanctuary for twelve years previously and had left on bad terms. To this day he hates Ralph. But Micki thought she might be able to help relieve Ralph of some of his bad publicity by leading him in the right direction.

At the time, Michelle Glean Simoneau was the sanctuary's marketing and public relations director. She was responsible for bringing in most of its money: calling donors for regular donations, and bigger donations in cases of emergency, and keeping on general good terms with the public. She excelled at her job and, over four years, built the sanctuary back up to near its former glory. Michelle's daughter even came to work with her, in keeping with the family business spirit of the sanctuary. According to the Vaughans, Ralph promised Michelle she'd be executive director when he retired, and she believed him. It was what she'd been working toward.

But "Micki wanted all the power," Greg said. "She wanted to have full control, boss over everything. And Ralph let her have it because it's his first cousin."

In 2011, after four years working alongside Micki, Michelle quit and Micki took over her public relations responsibilities. Ralph gave Micki the new title of operations manager. Shortly thereafter, Michelle told News Channel 8 that she left "because of conflicts over financial accountability."

"I would say we need to use this money for what it was intended for or we're going to get in trouble," she said, referring to a $300,000 donation she'd arranged, which she said came with stipulations for its use. "Finally, I had to leave."

Her daughter left, too; she was weeks behind in pay.

Kellie shook her head mournfully. "Micki hired eight or nine more employees, including her son."

"Who was the most useless piece of shit in the world," Greg added.

"Yep," Kellie agreed.

The following year, the *Tampa Bay Times* reported the sanctuary was under investigation by the Department of Labor for possible violations of the Fair Labor Standards Act, which covers minimum wage, overtime, child labor, and record keeping, and that the sanctuary had allowed employees' health insurance to lapse without telling them. Micki insisted the issue had been resolved and denied any wrongdoing. Then the payroll tax scandal happened. Then the workers' comp scandal. Then Ralph was caught stealing from the donation boxes. Meanwhile, more and more employees were finding themselves without paychecks. Kellie began working full-time at a nursing home to make ends meet as her full-time pay at the sanctuary dried up, until Ralph's sons hired her to care for Mrs. Heath.

"Mrs. Heath, to me, was like my grandma," Kellie said. "I lived with her for two years. Her and I were so close."

"That's what kept us there," said Greg.

"Micki always told us, since I was working and making money taking care of Mrs. Heath, that we didn't need no check," Kellie said.

Micki promised they'd be paid the next week, and the next—for seven months, they said, she paid one of them or the other. They never saw that back pay, nor the rest of the money they gave to help the sanctuary through what they saw were hard times. By the time Micki left, they said, they'd given $60,000 of their savings and were living on credit cards. They were lucky to have a good landlord.

"He never said a word about it," said Greg.

"Never said a word," Kellie said.

The Vaughans' deeper motivations for depleting their personal savings to help the sanctuary when they weren't getting paid regularly were still hazy to me. I suspected it was connected somehow to their feelings about Micki, but they never said so outright.

Then, in March 2013, Greg showed up to work at six a.m. as usual. He went about his business, checking on the birds, preparing their first meals. Everyone else would come in at seven, he thought. But seven rolled around and nobody showed. Then 7:15, then 7:30.

"At seven thirty, I thought something was wrong," he said. "I went in, there was keys everywhere, Post-it notes everywhere. Everybody quit."

"We're the only two that stayed," Kellie said, forgetting Chris.

Two months later, Micki quit, and Kellie went to the office preparing to take over.

"My first couple months, you can ask him, I was very overwhelmed," she said. "It was just like, 'We'll never get out of this mess. We're freaking screwed. We'll never get out of it.' Greg was like, 'Calm down, Kellie, you'll figure it out.'"

They were $24,000 in the hole at the bank, said Kellie. There were hundreds of thousands of dollars in unpaid bills stacked high on her new desk. They owed $8,000 to Office Depot alone.

"Now, how do you get away with it with somebody like that?" Greg laughed.

According to Kellie, the sanctuary's accountant discovered Micki had been turning in receipts every week for tanks of gas and cartons of cigarettes, claiming them as travel expenses. But more important was what happened when Kellie first went to the bank. There, she said, she discovered a bank account holding over $200,000, which, it seemed, nobody else had known existed. Kellie asked the bank teller what it was for and he promised to look into it over the weekend. When she came back on Monday, he'd resigned, and the bank account couldn't be verified.

"Did the teller just not want to be involved?" I asked.

"I'd say he was in on it," Greg said. "Because we never did find out nothing about it."

"We were never able to verify another account," Ralph told me. We were sitting opposite each other at a Greek restaurant across the street from a demolished shopping mall in Seminole. Three months had passed since our first conversation, and, after a stint back in Brooklyn, I'd returned to Florida looking for answers to the many questions that remained. Greg and Kellie had quit the sanctuary a month after I'd left—supposedly for health reasons, though Greg had told me in the garage that day that they were prepared to quit if Ralph didn't leave. They'd moved back to the midwest, where they still owned a house.

I'd discovered this after emailing Kellie at the sanctuary to request corroborating evidence for their claims of embezzlement—I was still missing important details, such as the name under which the account had been registered. Instead I received a response from Adrianne. She'd taken over as office manager, she said, and would not comment on Micki—who, I should note, was no longer associated with the sanctuary. Nor would she tell me whether Ralph was still director. After repeated attempts to get Greg and Kellie to corroborate their accusation of embezzlement, and Greg's repeated promises that Kellie would respond to my email tonight, tonight, tonight, he had stopped responding, and I'd boarded an airplane south.

"But the money went somewhere?" I asked Ralph at the Greek restaurant.

"We never had any proof that it was in a particular account," Ralph said. "We could never find any accounts where it was."

"So, this person arrived at the bank," I said, "after this other person quit, and saw something unusual . . ." I waited for him to finish my thought, but his eyes had glazed over. "What exactly did they see?"

"We were never able to find that supposed account," he said.

"So what was unusual?"

"We were never able to find an account with money in it, is what I'm trying to tell you."

"I know, but . . ." I was annoyed. He was sidestepping the question or he didn't understand it, and I couldn't tell which.

I tried again: "On the first day when this person went to the bank," I said, "the first day after Micki quit, they go to the bank and something is unusual. What is unusual?"

"I don't know. I wasn't handling the banking or the business there," Ralph concluded.

I took a deep breath and forked my spinach pie. I couldn't fathom the director of any business having no idea what was going on with the business. Aside from his ignorance concerning the goings-on in the office, it wasn't as if he was working every day on the sanctuary's birds, either. In a month and a half of working there, I'd seen him maybe a handful of times. Especially in the face of alleged embezzlement, what was Ralph doing to solve the mystery?

"We were never able to verify that account," he said to himself.

"Because the teller quit over the weekend?"

"Something like that. We were never able to verify or get any proof that anything was missing like that."

"I see," I said, though I didn't, exactly.

An old woman watched us from a table in the corner. Ralph had stopped to talk to her on the way in, leaving me waiting at the booth. He later explained she had donated many movable cages to the sanctuary over the years, by which I assumed he meant his warehouse. There were no movable cages at the sanctuary that I had ever seen.

"Ralph, when I was here a few months ago, your sons were in town, and they were asking you to sign something that would basically nullify your association with the sanctuary."

"Oh, no, that was . . ." He paused. "They bought the mortgage and took it over."

Four years ago, I thought.

"But I also know that what's been said about you in the media has impacted donations," I said, "and more recently, your sons and others at the sanctuary decided that your further association with the sanctuary would be harmful."

"Well, see, that's because people were believing the lies and not the truth."

I recalled what Greg and Kellie had said: that Ralph refuses to believe what people say about him is true. In the garage, I had asked them whether they truly believed he'd sign the paperwork when the time came. They told me that Beatrice and Ralph's sons had given him two weeks to move out of his house and into the warehouse. But as I sat across from him in this Greek restaurant, with its tinny ethnic music and its overhead lights bright as the sun, and my spinach pie growing cold upon its plate, I listened to him say that he was not yet ready to move, now two months later; that he had to make sure his birds had everything they needed, be it food or medication. He had to make sure the warehouse was ready for them.

"And you're continuing to work with the sanctuary?" I asked Ralph.

"Oh, yeah," he said, coughing. "I'm not giving up on working with the birds."

"But is there any way to change the minds of the public when they've read so much about you?" I asked. "There's no way for you to say, 'Hey, that won't happen again. We're doing things honestly now'?"

"Well, see, that's the problem when people either do the wrong thing or lie about you. The media, as you well know, likes sensationalism."

At this point, Ralph excused himself and ran out to his truck. He returned with a folded-up letter, which he handed to me. It was written by Kevin S. Doty, PA, whom Ralph explained is a friend of his, in addition to being a highly respected lawyer, who has argued cases in front of the Florida Supreme Court. The letter was dated a week prior and addressed to no one in particular, and vouched for Ralph's character, condemning "some people" who, "for whatever reason," have sought to trash the sanctuary's and Ralph's reputations.

Doty writes, "I know that because of my inside knowledge, that people have given the media false information and the media has printed or aired it even when Ralph tried to explain the truth." The letter doesn't specify what the false information is, nor who printed it. "It is very hard to defend oneself after something has already been printed," it continues. "I firmly believe that someone wants to get rid of Ralph and the birds for reasons that are not clear but appear to be nefarious."

Ralph offered to let me keep the letter.

"A major, major attorney for a newspaper years ago said, 'You can't fight people with either the microphone or the ink,'" Ralph said. "In other words, you can't fight people who have either the microphone or the ink." He coughed into his hand.

I arrived at Micki Eslick's South Tampa home that same sweltering day, long after sunset. We drank sweet tea out of plastic cups and chain-smoked cigarettes on beach chairs under a string of Christmas lights in her open garage. On a shelf behind me sat an urn containing Micki's mother's cremains. She died in January 2013, four months before

Micki left the sanctuary. Micki's friend paid $3,000 for the cremation, as Micki and Slick couldn't afford a burial. Later that year, they filed for Chapter 13 bankruptcy.

"I didn't do any embezzling." Micki laughed. "I've heard everything that I've done—supposedly done. Micki never embezzled any money. Greg was always in there when I counted it." She sipped her sweet tea and set it on the patio table between us. "The only money I ever took out of there was forty dollars to buy postage stamps. What am I going to embezzle? Do I look like I embezzled money?"

I'd come prepared to distrust Micki. In the car beforehand, I'd made a list of every question I needed her to answer. Why was she paid when no one else was? What about the receipts for gas and cigarettes? What about the bank account, the one that had disappeared?

She answered the door warmly and apologized for the state of her home as I entered. They'd been flooded a few weeks back during a big storm and were still in the process of repairing the damage. Their single-story, split-level house is carpeted in white and darkly furnished. A TV was turned down low in the den. Slick was out of town on business, Micki said, and she had just come home from her new job. I followed her through the half-lit kitchen back out to the garage.

Ralph and Micki grew up together; they were much like siblings, always very close. Before Ralph was born, his father had bought a Spanish-style mansion in Tampa, a sprawling estate, and housed his practice in the front, Micki's family in the back. All across the grounds, there were wild animals—turtles and snakes, birds and opossums—that Ralph Senior had healed and kept for his son. He built enclosures to protect the animals at night and heated pools for the turtles so they wouldn't be cold in the winter. Micki loved the animals and loved where she lived. Even so, she and Ralph had very different childhoods in terms of their privilege. She wasn't raised like Ralph. Her parents didn't have the kind of money his parents had.

"Why don't you go ask Ralph how much money he's embezzled from the sanctuary?" she said. In two and a half years, no one had asked for her side of the story. She kept a manila folder full of documents, which she agreed to show me before I left, including lists of sanctuary expenses she had charged to her personal credit card, totaling $6,000 for which she has never been repaid—the reason for the bankruptcy. Just after quitting, she gave copies of the papers to Kellie to fax to Ralph's sons. She doesn't know if Kellie sent the papers. She has never heard anything about them.

"How much money he's got sitting in the warehouse from all these donation boxes he's collected and not given the sanctuary one red dime," she continued. "You want me to get pissed? I'll get pissed."

I said that I was fine with her being pissed.

"Bullshit. Let me keep on going, what he's done. He couldn't sell the fucking boat that's sitting over there. Ralph thinks that Egypt or whatever country is going to give him a bazillion dollars—"

"The United Arab Emirates?"

"Whatever the hell it is anymore, and Omar's going to get it, and they're going to ship it off to Switzerland. They're going to put a heliport on it and redo it. And it's sitting over there rotten, covered in bird shit, and bees, and mildew, and everything else."

Things were coming full circle.

"He's been waiting to get that money now for thirty years, from Omar," she said, referring to Omar Bsaies, a sanctuary board member and former US diplomat now living in the Philippines.

According to Micki, not only had Ralph's relationship with

Omar never come to fruition financially, it had also soured other sources of income, including an impending contribution from the "Pog guy." The Pog guy was the son of the maker of Pogs, those cardboard collectible toys of the '90s. He'd arranged with his father to make a large donation to the sanctuary at a time when the sanctuary needed serious help. Ralph attended the meeting with Micki, Donnie, and Lisa, Ralph's drug-addicted, on-again, off-again girlfriend.

"We were just keeping our mouth shut, hoping Ralph wouldn't start on this overseas bullshit," Micki told me. But of course, he started talking. He told the Pog guy about the millions of dollars he was going to get from overseas, and his big plans for it. He was going to build a zoological park, he said. It was going to be incredible.

"The Pog guy says, 'Looks like you don't need my help.'" Micki scoffed. She'd witnessed this over and over. "That's what Ralph does with everybody," she said. "That's why the donations dropped off. Because everybody thought, 'Well, if he's going to get all this money for the park, why do I need to donate?' We didn't have any money."

Jimbo told me something similar later on. I had called him to check in. "Every time Eddie gets an opportunity to talk to a donor on his own," Jimbo said, referring to Eddie, sanctuary's new general manager, "somehow Ralph's psyche wakes up and he runs over and kind of ruins it."

I asked what kind of things Ralph would say.

"'We've got this under control!'" Jimbo said. "'A couple hundred thousand more dollars and we'll be back in business!' That's the best I can lie for him."

This wasn't a new problem. In 1986, a woman named Anne Bywater had inherited four hundred acres of land in the Hudson area from her recently deceased mother. The land was mostly underwater and was environmentally protected—there was a pair of bald eagles nesting in one of its trees—so Anne and her accountant had decided that, rather than sell the land, it would be beneficial to make a charitable donation of it. She chose to donate it to the Suncoast Seabird Sanctuary.

I had read that Ralph planned to turn the land into a zoological park like the one he'd wanted to build in the Gateway area back in the '70s. But in 2002, the *Sarasota Herald-Tribune* reported he was selling the property and that his plans for the zoological park had been annihilated by the no-name storm of 1993. This seemed strange to me—the no-name storm was bad, but it wasn't as if the park had been already under construction.

Anne had placed the land in a trust that was handled by Sun Bank of St. Petersburg. Carolyn Wiggins was the trust administrator. I asked Carolyn what she remembered of the deal. She answered glibly, "I honestly feel that Ralph Heath always had grand ideas that he never carried through."

I asked Micki if she remembered the plans for the park. "My husband was working there then," Micki said. "Ralph wanted to make another sanctuary, for an awful lot of money. Ralph would start, 'What I'm gonna do. I can do anything I want to. I'm Ralph Heath.' It just went away. You don't do business like that."

Around the same time, the board decided that selling Ralph's house would be in the sanctuary's best interest. They needed money, and after all, his house was sanctuary property. At first, Ralph agreed to sell it, just as he had the Stingray. But when the man showed up to install the "For Sale" sign, Ralph stopped him, again, just as he had with the car. And he could do that, because he was on the sanctuary's board and he had 51 percent of the vote.

His mother also sat on the board until she died. And Omar

Bsaies. And Jerry Allen, Ralph's close friend who died in 2013. And George Marler, a wildlife photographer who lives in Texas. I emailed George to see what he thought of the current situation. "If you know the history of the sanctuary then you should understand why Ralph will never leave," he responded. "I have never met anyone more passionate about wildlife, particularly birds, than Ralph . . . I have been with Ralph on many occasions over the past thirty-five years and he has always held the value of wildlife above that of money. Money to him, in my opinion, relates to what it takes to keep the sanctuary alive and well."

"They were yes-men for Ralph to do whatever he wanted to do," Micki said. "He's not going to sell anything. It's his. He thinks—maybe not so much now, but at that time, he thought the sanctuary and everything was his. Not that it was a nonprofit, not that 'it would go to the sanctuary.' His personally. He's always thought that. Always."

I asked if she remembered when Ralph began hoarding. The question confused her. "His whole life," she replied.

Ralph's father was one of the wealthiest doctors Tampa has ever seen. As Ralph put it, "money was literally no object." Growing up, Ralph didn't play with plastic toys, he said; he had the "real things." When he was ten, he got his captain's license—so his father bought him a boat. When he was fourteen, he owned eighteen cars. When Ralph entered high school, his father bought him a '61 Corvette. When he graduated, his father offered to buy him a new one. The '63 Stingray had just entered Florida; it was the first fuel-injected Corvette to enter the state of Florida. When Ralph told his father to trade the '61 for the '63, his father suggested he keep both.

"Then you have the '61 and the '63 to go along with your other eighteen or twenty cars," Ralph explained at the Greek restaurant.

But he didn't keep both cars. He insisted his father trade one for the other.

"I was trying to be nice on his finances," he said. "I made one of the biggest mistakes of my life."

The Stingray is today the "only triple red un-restored and all original 1963 Corvette of its kind" and is "irreplaceable"—it's the same car that, in 2014, the *Tampa Bay Times* reported Ralph sold to two different buyers, both of whom were now suing him.

"One of the worst mistakes, really?" I asked.

"That pair would have probably been worth a million dollars." Ralph laughed. "I mean any collector would have given me a million dollars for that pair of cars, instantly."

He continued: Because he was carrying so many classes toward his premed zoology degree, he'd spent seven years finishing his undergraduate studies. When he graduated, he still didn't know what he wanted to do. With his degree he could have gone on to become a veterinarian, or an animal researcher, or a dentist, or a medical doctor like his father. There was no rush to make a decision, though; Ralph Sr. had always told Ralph to get his education out of the way first, because once he had it nobody could take it away from him. Application of the degree could come later.

When Ralph graduated, his father said he should do whatever he wanted to do so long as he did it well. So, Ralph spent some time living at home with his parents to discover his calling. He enjoyed the beach-bum lifestyle and picked up a hobby making lamps out of driftwood to sell to tourists. But to a man like Ralph, the unglamorous life of an artisan wasn't enough to keep him motivated. He needed a profession better suited to his skills, something that demanded a rigorous capacity for reason, and had the potential for fame—and supporting his upper-class lifestyle.

One day, during this period of drift, he left the house to do some shopping. It was just before Christmas 1970. He was in the car with Linda, whom he'd married just out of high school, and would divorce three years after founding the sanctuary. On their way home, they saw the cormorant dragging a wing that would cause Ralph to abandon the lamp business for that of bird rescuing. Though, like the Stingray, he still keeps the driftwood lamps in his warehouse. He keeps everything in his warehouse.

"I felt that because my upbringing was so privileged," Ralph told me. "I said, 'Well, I'm going to dedicate my life to those who are less fortunate, in the bird category.'"

The rest is now history.

The first time I interviewed Ralph, I asked him what he's learned over all his years working with birds. "Probably the interesting thing that I have learned about birds by working on them so long," he told me, "is that they're an animal, or they . . . I guess I shouldn't call them an animal—they're a wildlife object that, you know, take a long time to bond to you. In other words, birds take a long time to bond to you as opposed to, say, a dog or a cat. And the interesting thing about birds is that once they bond to you, it's like a permanent bond. In other words, they'll love you for the rest of their life."

Birds he's known since they hatched, that are sick, will wait until he's home to die in his hands. If he's sitting in his bedroom doing paperwork, certain pigeons and doves that are free-flying in his house will come and sit in bed with him.

Jimbo sees Ralph's relationship with animals in a slightly different light. On the phone he told me, "Every day that I wake up in that house, I have to swipe away some spider webs to get out of my room. I'm not allowed to clean them because we don't have a proper home for the 'little spideys.' It's terrible. It's a terrible thing."

Jimbo had been living in Ralph's house for five months at that point, helping him pack for the move to the warehouse. He had packed everything but the birds and had done his best to clean the house and the yard as far as Ralph would let him. Ralph's sons had paid him to clear out a space in the warehouse in which Ralph could live comfortably. But as soon as Jimbo had cleared some room, Ralph filled it up again with furniture picked from the side of the road.

According to Jimbo, the warehouse is piled high with furniture, some roadside finds and some inherited. Micki had told me that among the furniture in the warehouse is some taken from an apartment that once belonged to Tara Gallagher, a former girlfriend of Ralph's and employee of the sanctuary who died in a drunken accident in 2010. It's not uncommon for Ralph to forage in the sanctuary's trash cans late at night, looking for things to salvage.

"'We can save that broken rattan sofa and one day we'll need it for one of our guests to sleep on!'" Jimbo said, imitating Ralph. "It's so beyond full, it's ridiculous."

He paused for a moment. "His heart is pure," he said finally, beginning to cry. "His thoughts about what is in the best bird interest is pure. He's still the best on-the-spot bird doctor there is. He's got these weird potions—who knows what's in it, but they fix everybody. He's fucking amazing."

I asked Jimbo whether anyone had thought to consult a psychiatrist. Ralph refuses, he said. "He will not for a moment adhere to the idea that he has a problem. No. He don't get it."

I asked if he was concerned about what would happen to Ralph once he moved into the warehouse. This is the place where he indulges himself, after all, and the results weren't exactly healthy.

"No," he said. He wasn't worried. "Ralph's going to die from whatever he dies from . . . in any place."

Spending enough time with Ralph, one will notice that he coughs a lot, but he doesn't smoke. In fact, he hates smoking—he told me this many times. When he was six, his father showed him what happens to smokers.

"Those were the days when a doctor could walk out of a hospital with any body part he wanted," he told me on the phone. We spoke for thirty minutes as I sat outside a café in my father's car. Fifteen minutes after we hung up, Ralph called me back to tell me the story of Grungy, the orphaned pigeon who saved his life one night. Then he segued into this one about his father.

Ralph Sr. died in 1986. He was something of a legend around Tampa, if one believes his son. He hovers in the background of the history of the sanctuary like a ghost; stalks around the perimeter, quietly watching, silently approving of Ralph's work. Ralph Jr. still has turtles that his father helped him put back together fifty years ago. Hearing Ralph talk about him is like hearing a Bruce Springsteen fan letter read aloud.

Once, Ralph's father had come home with a package wrapped in newspaper. He laid it on the walkway in front of Ralph's childhood home, now the sanctuary office. He told Ralph to unwrap the package. Ralph obeyed.

"And okay," Ralph said. "It's half of a human lung."

Ralph tells this story with great excitement and a certain filial pride. He has a knack for telling the same stories over and over with equal enthusiasm. He has many stories about his father.

Ralph's father handed him a scalpel, and Ralph opened the lung.

"The only way I can describe it to you," he told me, "it looked like an old Georgia asphalt road inside."

He has never tried a cigarette.

Birds, especially pigeons, are linked to a slew of human illnesses. Most of them affect the respiratory system. Bird fancier's lung is one of the most common. It's associated with long-term or intense exposure to bird feathers and droppings. It causes what's described as a "ground glass" appearance of the lungs so that they appear jagged and milky on X-rays. It also causes shortness of breath and a dry cough. Other bird-linked illnesses, such as cryptococcal meningitis, affect the brain. It was thought that Micki Eslick's husband, Slick, had early-onset Alzheimer's until a spinal tap showed it wasn't that, but rather cryptococcal meningitis. He can no longer spend any time around birds.

In August 2012, Robin Vergara, a sanctuary employee, sent an email to Ralph's sons that warned them of the trouble the sanctuary was facing. "Birds are dying at an alarming rate," it read. "Bird enclosures are decaying at a fast rate and there are no funds for repair." It spoke of a lack of trust between staff and management, and of the increasing difficulty of maintaining donors.

Attached to the email was a video outlining Robin's action plan for repairing the sanctuary, "Wings of Change: A Win-Win for All." He had worked with Philadelphia-based marketing guru and radio personality Mel "Toxic" Taylor to come up with the plan. Among its many points was the necessity of finding new management and clearing the sanctuary of bad press.

"We know hurricane season just started," Taylor says in "Wings of Change." "Extra warm waters this season. Are we ready to evacuate? That warehouse, is it available?" Earlier in the year, the *Tampa Bay Times* had reported that the sanctuary had received a $100,000 donation from the Embassy of Qatar to help construct hurricane resources. The sanctuary could have used the donated money to buy generators for the warehouse and repair the damage to its structure. Instead, Micki informed Robin via email that there was no warehouse that could accommodate birds in the case of an emergency. "At this time we don't have a warehouse to house the birds and that is something that Ralph is working on," she wrote. A new warehouse never materialized, nor did space in the existing warehouse, nor the generators, nor a plan for evacuation—nor an accounting of how the money was used. It seemed to have just disappeared.

"If a major storm would have hit that summer, it would have been a disaster like no other," Robin told me via email. Among the concerns listed in his email to Ralph's sons, he included this lack of an evacuation plan. Ralph's sons never responded.

Fed up, Robin finally left to open his own bird sanctuary, the Gulf Coast Bird Rescue, in partnership with the county and the school board—partnerships that weren't easy to secure, according to him. It took a lot of time and effort, and careful planning, to convince those in power that the people behind Gulf Coast Bird Rescue were different from the people at the Suncoast Seabird Sanctuary. The scandals there had tainted the reputation of bird rescuers in general.

In April 2014, Florida Fish and Wildlife carried out an inspection of the Suncoast Seabird Sanctuary, Ralph's house, and the warehouse, responding to complaints that his animals were living in squalor. What they found were birds "confined in unsanitary conditions" and injured wildlife that had not received care. They issued fifty-four violations for Ralph's failure to maintain daily treatment and feeding logs for his birds. They issued two violations for rehabilitating migratory birds at an unapproved location. They issued one for failure to maintain daily clean water for seventy-eight turtles housed in a single unsanitary pool smelling of ammonia. And one more, for failure to meet minimum caging and water requirements for an African spurred tortoise. Wildlife veterinarians called in to assess the situation found ten birds with extensive untreated injuries. Eight of them had to be euthanized for necrotic flesh and exposed bones. The others were taken to outside veterinarians.

Less than two months later, wildlife inspector Lar Gregory returned for a follow-up visit. Conditions had improved to a degree. He reported that Ralph was "no longer rehabbing birds at his home or the warehouse," and that the warehouse had been outfitted with two ventilating fans for the turtle pools. The African spurred tortoise had immediately been removed. The pigeons in the warehouse were housed in a bigger room, away from the turtles, and appeared to be fine.

I asked Gregory why Ralph's rehabber's license hadn't been revoked the first time they inspected. "These people that do rehabs, other than donations, they don't really get anything for it," he told me. "A lot of these people have their heart in the right place and they want to help animals, and we like to see that. So, we really just want to see them in compliance. We're not in the business of shutting people down." Florida Fish and Wildlife doesn't regulate pigeons, as they're not considered wild—they're considered domestic animals. There is no regulation on the number of pigeons a person can have.

I asked what would happen to the pigeons in Ralph's house and the warehouse if Ralph had to be shut down.

"You'd have to consider euthanasia," Gregory said. "Because you can't just release those out into the wild."

In January 2016, Jimbo told me, Andrew, Alex, and Peter planned to return to Florida to force Ralph out of his house and into the warehouse. In the Greek restaurant, I had asked Ralph why he'd want to live in the warehouse as opposed to moving into a new house. "Well, one thing that I've always worried about is hurricanes," he said. "I've always been worried about hurricanes because, actually, the last big hurricane that hit here was 1848. So, we're way overdue for a major hit."

Ralph had spent a good 50 percent of our second phone conversation talking about hurricanes. He has an encyclopedic memory for them, and a visceral fear not just for his own life, but also for the lives of his birds. He had outlined this horror for me: of a bird lying somewhere without his knowledge and him unable to help it. Recalling the rescue of an orphaned baby pigeon later in our conversation at the restaurant, he began to cry openly. I reached over to the nearest table to find him a napkin. A hurricane scenario is the worst he can imagine. We'd been discussing the stunning number of pigeon species—there are over nine hundred varieties—but we were off on a serious tangent now.

Ralph repeats himself a lot while he's talking, as if he's lost track of the thread of his thought and has to start over to retrace it. Consequently, it takes him a long time to tell a single simple story, and he repeats the same stories over and over again, almost verbatim.

"Everyone laughed at me about being afraid of hurricanes until Andrew hit Miami," he said. "Then they weren't laughing anymore, when the place looked like a nuclear bomb went off. That storm that hit Mexico, did you know it went from a Force One to a Force Five in twenty-four hours?"

I said that I did.

"That was unheard of. And also Andrew, the one that hit Miami, it increased unexpectedly right before it hit shore. And Charlie, that was coming up the west coast, that we thought was going to hit here, the abrupt right turn into Charlotte Harbor, it was increasing slowly, but they said if it hit—look up the history—Tampa Bay was bull's-eye. I mean, they were expecting it to come right up the coast and then turn in, right into Tampa Bay. And every single—what's the word I'm looking for?—seasoned weather person in this area had said, if it made it to Tampa Bay, it would have been a Force Five. The computer graphics, printouts, predictions, whatever you want to call it, if it had made it to Tampa Bay, it would have been a Force Five."

I searched for the proper way to respond. I questioned how helpful this talk about storms was. To me it seemed, all of it, very symbolic. Some storm or destructive force appeared to be churning just offshore of Ralph's psyche, threatening to tear apart his fragile ecosystem.

"I bet it feels nice to have the birds in the warehouse," I said. "So, not just for your own protection, but for the birds' protection as well."

"I couldn't stand it if my birds weren't okay," he said. "That I couldn't take, you know?"

I said I believed him.

"When you've raised a baby bird from—like Grungy, for instance—you live with them and stuff like that, you don't want them to get killed. And people just flat don't realize—people don't understand how powerful you have to build a building to withstand hundred-fifty-mile-per-hour-or-more winds. It's just absolutely crazy.

"And actually, Chief Jim Billie, chief of the Seminoles, he told me stories, eons ago, before the weather station or people keeping the weather reports or records, of what they called the 'bad winds.'"

In the moment, this talk felt normal. We sat in that restaurant for three hours total, while Ralph did most of the talking. The waitress reminded us twice to pay at the front. Ralph asked for a to-go container to bring crumbs from my spinach pie and scraps from his salad back to his chickens. Ralph has fifty chickens in the warehouse. He feeds their eggs to his crows.

Ralph is one of these people who doesn't need much prompting in interviews. Our conversation wended its way from the evils of mass media, to hurricanes, to pigeons, to his contempt for the condos next door to the sanctuary, to the fate of the environment, to the fate of the human race—and finally to the night before, when he saved a starling stuck to a phone pole. A friend had called him on behalf of another friend. Ralph had wrested the bird free, and was keeping it in his house now, until it got its strength back.

"The birds are lucky to have you," I said.

"Oh, I like my little creatures," he said. "The most important thing is—and you're welcome to put it down even though it forms more of a psychic story—is the number of herons, egrets, pelicans, even the lowly gray pigeons that will wait—usually you could almost set your clock, it's usually between eleven and twelve o'clock at night, and it's only when I'm alone, and they will either fly, walk, or crawl up to me, or even land on me in some cases, and not just here but in some places when I'm away from the sanctuary."

"Like they're sending messages to each other in other places?"

"No," he said. "Actually, I asked Chief Jim. I always refer to him because he's forgotten more about the wildlife and the environment, especially in the Everglades, than most anybody ever knows, and things that aren't in any books."

Chief Jim Billie is the longest-serving chairman of the Seminole tribe of Florida. Jim's name resurfaced repeatedly while we talked. Their meeting had become a pearl in Ralph's mind, a grain of sand around which Ralph had deposited layers and layers of soft tissue that had hardened.

"One time I said, 'Chief Jim, why are these totally wild birds that I haven't, so far as I know, had any contact with before walk—

I'm talking about herons hobbling up to me with broken legs—walking up to me with broken wings, pigeons landing on me that are sick or injured?' He says, 'What you're doing is putting out an aura like a radar.' And he says, 'The birds are picking up on it.' All species, because I've had pelicans and egrets. I've had birds hobble out from under cars and everything, and come up to me, when they know I'm there at night. It's usually between eleven and twelve o'clock at night. It's not in the daytime, it's usually at night. He says, 'You're putting up this aura, like a radar, that the birds are picking up on. They're coming to you, to know that they can get help.' And I'm talking about, you get a wild heron or egret just to walk up to you, stand there without moving, and let you pick it up, and operate on it without any anesthesia. I'm talking about setting wings, setting legs, sewing them up. Now, because it's just me and it's usually late at night, so nobody else is around. The bird will lie absolutely still, just like that dish there. And looking at me. These are totally awake, alert birds. You go pick up a blue heron and get him to lie on his side, not moving, with no anesthesia."

The story he'd been trying to arrive at finally came out: Late one night, between eleven and twelve o'clock, a blue heron approached Ralph at the sanctuary. He was alone. It walked directly up to him and stopped, like a lady wanting to dance. Instinctively, Ralph wrapped his arm around the bird, and four of his fingers slipped into a large wound, still very fresh. Removing his hand, he found it was covered with blood. He brought the bird to his house and laid it across his washer and dryer.

"I mean, this bird's not even moving," he said. "He's letting me do whatever I want to with him. I laid him down and turned him over. You gotta remember: this is a totally alert, alive, awake, nonanesthetized heron. He's not sedated in any way. I laid him on his side, and he let me do whatever I want to with him. So, I turn him over. And he's looking at me, you know? I saw the gash in his back and it was easy to see what it was, because it was a fishhook at the bottom of this gash. The hook wasn't very big to make a gash this big and deep, and it was right along the edge of his backbone."

Ralph stitched the bird up, using skills his father taught him.

"I always keep sutures around with me. I started sewing and sewing and sewing and sewing. It's sew and tie it off, come a little bit further and sew and tie it off. I did that because I didn't want to take a chance on him turning around and grabbing one of the stitches, and pulling the whole thing out like unraveling a sweater." He made a spiral with his finger. "You can open the whole damn thing back up. I did interrupted stitches on purpose, and then once I was finished, I coated the whole wound real thick with antibiotic ointment."

Ralph kept the heron in a cage on his dining room table.

"I would pick him up and put him on the floor while I changed his cage and everything. Then pick him up and put him back in. This was now, probably, three years ago—and I on purpose didn't take him over to the hospital because obviously a bird that trusts you so much to lie still on your washer and dryer while you sew him up—to say he trusts you would be an understatement. I mean, he's not moving! He's not moving a feather! You know? Not kicking or flapping his wings or nothing. He's just lying there absolutely like he's in a coma. But he's not; he's wide-awake looking at you."

When the bird had recuperated, Ralph brought it to his porch. He opened the door of its cage. It stepped out onto the sand and sat looking at the water. For a long time after that, if Ralph walked down on the beach at night, the bird would fly to him.

"You know how I knew it was the same bird?"

I asked him how.

"He had a big oily patch on his back. He had a big oily spot on his back where I coated him with the antibiotic ointment."

Ralph smiled. This was his favorite story.

In January 2016, *Tampa Bay Newspapers* ran an online story about a blue heron approaching Ralph at the sanctuary. It was late at night, and Ralph was there alone. Just as in the story he'd told me, the heron reportedly approached him and Ralph put his arm around it. Only, in this version of the story, there was no hole in the heron's back—this heron's leg was broken. It was dragging a foot.

Just as in the story he'd told me, Ralph said he realized the bird wasn't going to hurt him. He laid the heron on the examining table. He began to work on the heron alone.

But in this version, he put a splint on its leg and taped it in place. The bird watched him the whole time. It acted like it had known Ralph all its life.

Ralph kept the bird for a month, then set it free. It had come back to the sanctuary twice since then.

"We knew it was him because of the bump on his leg where the break had healed," he said in the article.

Near the end of our conversation at the Greek restaurant I had asked Ralph, for the third or fourth time, if he would show me the warehouse. Again he'd said no. He explained that it was "personal," that he "keeps it private," even though, until recently, it was owned by the sanctuary and by Florida state law any animals kept by the sanctuary as permanent residents must be made available for public viewing. Now housed in a space owned by his sons, the birds were officially Ralph's personal collection. Different laws applied to them.

Ralph's birds wouldn't react to me like they do to him, he explained. "Birds take a real long time to bond to you. To bond to you is like a permanent affection, or a permanent love, or a bond. Birds, the ones that I live with, are very different when there's a person around as compared to me."

In May 2016, Florida Fish and Wildlife received an anonymous report describing the abusive and disgusting living conditions of the animals kept in Ralph's warehouse. Lieutenant Steve Delacure and Officer Robert O'Horo investigated the claim. "A cloud of rancid-smelling particulate hung in the air," O'Horo reported. The warehouse was so dark he needed to use a flashlight. "As I entered the room, my boots sank into the wet spongy mix of substrate, feces, spoiled food and feathers. I observed roaches running over my boots on several occasions and my boots were caked with fecal material after my visit."

The report included photographs. I finally had my access to the warehouse. There was the sprawling, windowless building, light blue with brick borders and painted details of palm leaves. Inside, turtles with deformed shells swam in murky, fetid pools, and crawled around on filthy concrete floors. Ducks and chickens wandered in dark rooms with walls coated in mold and waste. Pigeons and exotic birds sat locked in cages caked with droppings and rotten seed. The wing of a federally protected laughing gull had been amputated and bound in duct tape. Some of the birds had gone blind. None of the animals had clean water.

Thirty-three box turtles were removed immediately, and nine red-eared sliders. O'Horo issued five misdemeanor citations for rehabilitating wildlife at an unapproved location, possession of migratory birds without a federal permit, failing to maintain wildlife in humane or sanitary conditions, failing to maintain intake records for the animals on-site, and possession of Florida box turtles without a permit.

Ralph's federal rehabilitator's license had already expired and will likely not be renewed. Lieutenant Delacure recommended a Pinellas County judge suspend or revoke Ralph's state license; he was still on probation for the fifty-nine violations Florida Fish and Wildlife had issued him in 2014, and was charged with violating his probation. He pled not guilty to all charges. I called him to ask what happened.

"Oh, I just hadn't had time to. My assistant who was working there passed away unexpectedly," he said. "And then the fellow who was helping me—you know Jimbo? You remember Jimbo, of course."

I said I did.

"He had to check himself into alcohol rehab or *he'd* have been dead. So, I kind of got behind on my work there and everything."

He told me he had a new person working with him now, a woman who was strong and very good with the animals. But he had a lot to do that day to prepare for the storm that was expected to make landfall Monday, he said. I should give him a holler the day after tomorrow.

"I just have one more question, Ralph," I said. "When we talked in November, you were preparing to move into the warehouse. Did that ever happen?"

"Right now I'm just trying to get organized," he said. "Call me back Tuesday." He hung up. That was that.

When I had returned to Brooklyn in November 2015, my husband and I had adopted a green-cheek conure and named her Magnolia. She was about the size of your hand, and jewel-toned, with green wings, bright blue flight feathers, and a smudge of a red heart on her breast. She belonged to a South African fashion model who traveled frequently for work and was concerned that Magnolia, née Kitten, wasn't getting the love she needed. Magnolia came to us in a stately metal cage with a scrolled hook at its top, food, bedding, and two dishes. She loved bananas. She made happy chirping noises and sang with the doorbell. She loved to bathe in the sink and ride around in my hair, and she preened constantly, as if invested in her appearance. When we left the room, she'd cry and call for us until we returned. During the daytime, she'd find a safe place beneath my ponytail and stay there for hours while I wrote, cooing softly, nuzzling in my neck. We awoke in the mornings to her singing in the sun. She learned games quickly. She always wanted to play.

But Magnolia was filthy. She shat all over the furniture, and us. Our attempts to return her to her cage ended with bloody fingers. She was moody and restless—full of energy, all of it focused in her beak, which she ground constantly. She chewed everything: papers, electrical cords, books, scabs. She sought out our wounds and pulled at them, then groomed the open flesh even wider. She had a taste for blood. She ate her own feces. When she was mad at us, she tantrummed in her cage and threw seed and bits of produce all over the floor. Reasoning with her was futile. I stroked her head sweetly, looked into her eyes: in them I saw the void of a mind I will never understand. Magnolia was an enigma. Stubborn, willful, animal, id. She derived pleasure from our frustration. To her, everything was a game.

On the fourth day of our cohabitation, I awoke to Magnolia's chirping. I had learned to anticipate it as a response to the alarm I set on my phone. I pretended to fall back asleep, not ready yet to entertain her. My husband and I were unaware we should set a schedule for cage time; I felt it was expected of me to take her out of her cage on my way to the bathroom. But that day, I didn't. I felt her eyes on the back of my head as I passed her cage, rounding the corner of the kitchen. Leaving the bathroom, I moved directly to the coffee maker. I carried the guilt with me through my morning routine, hearing her stir in her cage, vying for my attention. I remembered something Greg had told me, a rule of thumb for releasing birds: If it can fly, it's free. Like spirit, like wind.

That night, we returned her to the fashion model, who had given us time to decide. We'd realized we couldn't give Magnolia the attention she needed, either. In our four days as her caretakers, we had been forced to rearrange every aspect of our lives: putting off sex and socializing, smoking in the bathroom, going to bed early. We had thrown away all of our Teflon cookware. We had watched training videos on YouTube, joined a Facebook group for conure owners, ordered books, visited Reddit forums. I had asked my friend with a bird: What does it mean when she taps her beak this way? What does it mean when she bobs her head this way? We had lengthy discussions about the meanings of certain gestures. In the end, my husband and I decided she spoke a language we couldn't learn.

Rabbit

My grandmother is a sensible woman with expensive taste. Most of the gifts she's given me over the years have been jewelry. Or clothing: when I was young she loved to take me shopping at Neiman Marcus, then visit the top-floor restaurant for tea and miniature sandwiches. Each time I visited Cleveland, she took me to get my long, tangly hair cut, and for a manicure if I'd upheld my promise not to bite my fingernails since my last visit—not gifts so much as traditions. I'd sit nearby in the beauty parlor as my grandmother got her practical short hair restyled at her weekly appointment, then accompany her to her friend's home salon, where she'd get her acrylics touched up and I'd eye the wall of colors for one that fit my mood.

I was eight when she gave me the rabbit, an unusually sentimental gesture in our relationship. She made no overtures about it, simply handed it to me one afternoon as we were passing each other in the hallway. I was coming up from the basement after some ritual exploring of the terrain—old stacks of *Playboys*, boxes of liquor, and letters on brittle paper; she was hanging her coat in the entryway closet after running errands. The endowment happened, and then it was over.

The rabbit was gray with satin ears, a smooth pink nose, a white face and white hands and feet. Its long limbs made it perfect for sitting upright or pulling close, carrying by one arm or dressing in doll clothes. Though I felt I was too old for stuffed animals, I started sleeping with the rabbit every night. Eventually, her soft fur wore down and became patchy, her body flattened, and her white face grew yellow.

My grandmother met my grandfather in 1945 at a Valentine's Day dance held at the local synagogue. She was nineteen, just out of high school and planning to go to college for accounting. He was twenty-two and just out of the army, where he spent three years stationed in the Pacific theater and contracted dengue fever. This is one of my favorite stories: he spots her across the ballroom, sitting with her girlfriends. She's the most beautiful woman in the room—big brown eyes, strawberry blonde hair, serious expression. He works up the nerve, approaches her table, asks her to dance.

"No," she says. "You're too short."

My grandmother is a certain kind of independent. She was the bookkeeper for my grandfather's industrial-barrel business from the day they married until the day he died, sixty-five years later. She was also an expert homemaker: adept at making beaded floral arrangements, crocheted blankets, and ample baked goods. Though introverted, she's notorious for doing and saying whatever she pleases at any given moment. Once, in the middle of a dinner party, she removed a section of newspaper from her purse, opened it, and proceeded to read.

One afternoon, when I was visiting in the spring of 2011, my grandmother and I went out to get our hair done. Crossing the parking lot to the beauty parlor, I asked her what makes a person classy.

"The way they act," she said. "Whether they're kind and considerate. Not fake."

"Not fake?" I said. "So, if you can be kind and considerate—"

"But not two-faced," she said. "Saying one thing and doing something else."

I reflected on this for a moment as we continued across the parking lot, how true this was of the Gerard family. I thought of my grandfather, always keeping his word in business—and my grandmother, getting up at four o'clock every morning to make sandwiches for the men working in the barrel yard. Yes, my grandparents considered good manners and promise keeping to be simple matters of respect and forthrightness, I thought. Certainly this had been passed down to their children and their grandchildren. I swelled with pride.

"My hairdresser's lost two hundred pounds," said my grandmother, interrupting my train of thought. "And he's still fat."

While my grandfather was dying, my grandmother would curse the cancer eating away at invisible places inside him. Her inability to stop it infuriated her. Contrasted against the rest of the family's vociferousness, her anger was quiet, buried beneath the surface, and would come out in bursts: lashing out, then retreating immediately. Where the rest of the family argued and ordered one another around, she seethed. She drew herself inward. She talked less and less. Her mind seemed to wander. Days before my grandfather passed, while he lay on a hospice bed in their living room, she suggested we try to get him up and walk him around. It was inconceivable to her that he'd never again be the man who approached her at the Valentine's Day dance in 1945.

When he was a young man, my grandfather harbored dreams of being a writer. A chest of drawers in the basement held his screenplays and short stories—love stories, mostly. My grandfather was a hopeless romantic. Even into their final years together, sometimes he'd turn to me and say, about my grandmother standing near us, "Ain't she a foxy lady?" She would respond with, "Oh, shush."

He was an avid reader of novels. He loved mysteries, but hated violence. He was always singing songs by Frank Sinatra and Rosemary Clooney, and Michael Jackson and the Rolling Stones—anything that got his body moving. When he danced, his hands became pistol-shaped and he pointed them away from his body, winking at his dance partner. His specialty was speechifying—he was always ready to drop a nugget of wisdom, based in principles starting with God and descending to the bothersome chipmunks eating his flowers—but he wasn't arrogant. He was easily embarrassed.

"Most important," he'd always say, "guard your health. Without your health, you have nothing."

One of my earliest memories is with my grandfather at the beach. It was a rare visit to Florida, before he stopped flying altogether—he always got sick on airplanes. I was three, and was wearing my zebra-print bathing suit. My parents had warned me about the sand spurs on the dunes, but I wasn't listening, or didn't listen. I marched straight into them, and before I knew it, my feet were on fire. Sand spurs were stuck to the bottoms, and the more I tried to avoid them, the more they stuck. I started to wail. My grandfather ran over and lifted me out of the spurs, and carried me down to the shore to wash my feet in the water.

When the last of their three children had left home, my grandparents moved out of their house on Rochester Road in Shaker Heights and into a sprawling ranch house in Pepper Pike, another affluent Cleveland suburb. I spent weeks there when I wasn't in school, exploring the wonders of their dressers and record cabinets, the mysteries of my grandmother's home office, and the sprawling landscape of the property around their home. The land was partially wooded and populated with deer and chipmunks, skunks and fireflies: Ohio creatures that I never saw in Florida. I'd wander through the trees, hunting for animal bones and unusual seed pods and peering into the neighbor's yards. I spent whole afternoons rolling down the small hill between my grandparents' paving-stone patio and the bottom of their backyard. I never left without visiting the room in the basement containing select garments from my grandparents' old wardrobes—my grandfather's army jacket, my grandmother's knee-length rabbit fur coat—or reading again my grandfather's screenplays and short stories, and my aunt's letters from summer camp.

The house was also where I learned about my father as a child. I found his bar mitzvah photos in the guest bedroom nightstand; I ate the foods he ate growing up. Over a dinner of beet borscht in my grandparents' kitchen, I would listen to their stories about him taking ballroom dancing lessons, tripping on acid, teasing my aunt's boyfriends. Later, on the way to his favorite ice cream parlor, we'd drive by his elementary school and the field where he played little league baseball. I'd fall asleep that night beneath the same blankets that kept him warm when he was a child. My childhood was mapped onto his.

My father moved to Florida after college. His brother landed in Salt Lake City and his sister in California. Thanksgiving was our yearly reunion, and the family's pilgrimage back to Cleveland. The holiday followed a procedure: Two days in advance of the dinner, my grandmother would tell my father to bring the chairs up from the basement. Then would follow a day of relative calm, broken up by my grandparents shouting reminders at each other across the house. I would begin sneaking bites of the mandel bread my grandmother left wrapped up in the kitchen.

Dinner would finally come and midway through someone would tell a joke in poor taste, and we'd all exchange looks. My father's sister would breeze in with her family toward the end of the meal, looking frazzled, having caught a late flight. When they finished eating, we'd move to the yellow-lit den for drinks. I'd fall asleep with my head on my dad's lap.

My grandfather fought his cancer for years. As it advanced, his children spent more and more time in Cleveland. Near the end, my father drove there monthly to help him eat, bathe, and follow his treatment schedule. He spent weeks at a time away from my mother and from his work. His phone conversations were consumed by names of doctors and experimental procedures, medications and dietary restrictions. Though grateful for his children's help, my grandfather was a stubborn man. He grew depressed. His weight plummeted. He received a colostomy bag, and dressing himself became nearly impossible—not only because of the physical difficulty of it, but also because the humiliation of having to hide the bag was almost too much for him to bear. Once an avid moviegoer, he could no longer sit comfortably through a whole film. Unsteady on his feet, he couldn't walk to the mailbox at the end of the driveway.

One night while I was there, I heard my grandfather calling from the bathroom. My uncle was with me—we rushed in together. His colostomy stoma had opened, and his intestines were spilling into the plastic bag attached to his abdomen. We drove to the hospital, and my grandfather spent the next several hours awaiting surgery while his guts sat outside his body, next to him on the table, under a wet towel.

A few days later, we took him to the movies. He made it through the film, and on our way to leave the theater he stopped at the bathroom. We again heard him calling for assistance—my uncle went in. My grandfather's pant leg was spotted with urine; his embarrassment was huge. He couldn't stand the idea of walking to the entrance, of who might see him and what they'd think. He was almost in tears. My uncle tied a jacket around his waist.

In August 2012, my mother called to tell me my grandfather's body was shutting down. He would likely die by the end of Saturday; it was currently Friday morning. I flew to Cleveland, and spent the next two weeks with my family sitting vigil in my grandparents' living room, waiting for him to pass. His hospice bed was situated in front of the picture windows in which my grandmother kept her potted plants—rubber tree, philodendron, peace lily—overlooking the verdant backyard and the woods beyond it.

As the days dragged on, his speech became distant and confused. His breath rattled. He called out in his sleep for relatives long dead. Our family worked in rounds, keeping him dosed with pain medication, and to keep his mouth from drying out, swabbing with a pink sponge on the end of a plastic wand. He developed thrush in his throat: a white paste that spread across his palate and tongue. He fell into a delirious sleep from which he never awoke.

To keep him from developing sores, it became necessary to reach across his small body and move him. We cradled him toward us in the sheet while one of us stuffed a pillow underneath him. To keep him from aspirating his morphine, we tucked pillows around his head and placed the dropper inside his cheek, rubbing the outside gently to work the medicine into his bloodstream. We combed his hair because he would want us to keep him looking presentable, even in his state. We rubbed lotion into his hands. We took turns reading him books by Alexander McCall Smith, his favorite author.

One night, my father and I agreed to sleep in shifts so that one person could always be awake in case he needed to be made comfortable. It was already late when I fell asleep on one couch, my father sitting vigil on the other. At seven in the morning, I awoke to the sound of my father sipping coffee and realized he had never let me take my shift. He told me he'd wanted to let me sleep, but I was angry. "You denied me the chance to care for my grandfather," I told him. I felt betrayed. I felt he'd taken something from me that wasn't his to take.

In the years since then, I've wondered what my father was doing as I slept. I imagine he watched his father breathing. He held his hand, and fed him morphine. He mopped his mouth. He stared for a long time into the darkness. He rested his face in his palms and shook his head in bewilderment. He imagined my grandfather when he was young. He watched the sun come up.

My father and I were pallbearers. With white gloves, we bore my grandfather's weight from the funeral parlor to the hearse and from the hearse to the grave. As they lowered him into the ground, we dropped our gloves in after him. In the Jewish tradition we bury our own dead, so each of us shoveled a mound of earth from a bucket onto the casket. Inside, my grandfather lay with a bottle of Johnnie Walker Blue, his favorite. My grandmother sat on a folding chair at the end of the front row with an American flag on her lap folded into thirds.

Soon after his death, my grandmother moved into a tiny assisted-living apartment: kitchen and living room, bathroom, bedroom, walk-in closet. There, she set about learning a new routine. She ate meals in the downstairs restaurant with the other residents—including Thanksgiving lunch, before having dinner with the family. Once a week, she took a van to the beauty parlor, and to the library. She participated in group activities: macramé, card games, trips to the movies. She joined my parents' monthly book club. But for all the time she spent among others, she talked very little. She made few friends.

At the time, my father was semiretired and working on the novel he'd always wanted to write. Every morning, he'd bring a mug of coffee to his home office and think about plot. Around eleven o'clock, my grandmother would call to ask if he was coming to see her that day and he'd answer yes—then remind her, "Mom, I asked you not to call before three." That evening, he'd pick her up at her apartment and bring her back to the house, or take her out to dinner at her favorite family-style Greek restaurant, where she'd order snails. He'd drop her off at her apartment that night and stay for a while to make sure the new phone in her bedroom was working and that she was taking her medication, which she couldn't be trusted to do, even if she told him she was. He'd add books to her e-reader, then remind her again how to use it, and her cell phone. He'd kiss her on the top of her head. Then he'd leave.

My parents hosted the first Thanksgiving without my grandfather in 2013. My father's siblings flew to Florida with their spouses and children, and we gathered around a new table together, in a new house, and began the process of building a new tradition. Familiar faces were missing: my grandparents' closest friends, my dad's cousins and their families. But others, like my mother's sisters, had never been to a Thanksgiving in Cleveland. Dishes were passed around the table, wine bottles emptied, everyone laughed and stuffed their faces. In the midst of it, my grandmother fell quiet.

At the end of the night, my husband and I drove her back to the independent-living facility. Her apartment was cluttered, though a woman came once a week to clean it. Various papers were piled atop the kitchen counter with an open bag of individually wrapped Ghirardelli chocolates, two potted orchids, orange bottles holding pills, and orphaned items taken from the cabinets—cooking spray, cleaning supplies, an immersion blender. A cardboard box on the sofa held shiny blue gift bags, each containing a piece of costume jewelry she'd bought as Hanukkah presents from a vendor who made rounds in the facility. The side table held dishes of colored arts-and-crafts beads and a few lanyards. I asked her if she wanted help cleaning, but—as often happened by then—she didn't answer. We followed her to her bedroom, where her engagement photograph hung next to my grandfather's.

"He looked like a movie star," she said.

He did. Blue eyes, pouty lips, sensitive jaw, thoughtful expression. He was a man who took care with his appearance: fine clothing, trimmed hair, clean shave, shirts always pressed, hat always matching his coat. His good taste extended into choices of restaurants and Scotch, friends and films.

I reminded my grandmother of a conversation we'd had years before. I was a freshman in college and dating my first serious boyfriend. We were fighting nonstop—he was jealous; I was secretive. Neither of us was happy. One afternoon, after a particularly bad argument, I called to wish my grandparents a happy anniversary. I asked my grandmother how she managed to make a relationship work for almost sixty years when I couldn't even make one last six months.

"You said he was cute," I said.

"He was," she said. "Then he went and died on me."

THREE YEARS AFTER MY GRANDFATHER'S DEATH, my grandmother suffered a massive stroke. Though she had a DNR, she was resuscitated and awoke after the episode without the ability to walk, swallow, speak, or use a pen. The right side of her face hung slack from her cheekbone. Her right arm lay limp on the bed next to her, often contorted into a strange shape that she couldn't feel. In the hospital, she developed a urinary tract infection that rendered her temporarily catatonic. She was in the ICU for two weeks. All we could do was wait, and hope she recovered.

My grandmother now lives in Sabal Palms, a hospital facility for residents who need round-the-clock nursing staff. The door of her room is always open. The room is tile and partitioned off in the middle by a curtain separating her space from her neighbor's. She isn't allowed to drink anything because she might aspirate it, but she begs for water constantly in words often difficult to understand. In place of water, she's given lemon-flavored frozen suckers. As needed, an aide connects a saline pouch to a tube in her belly. She practices swallowing with pudding. When she needs to use the restroom, a complicated machine is wheeled down the hallway to lift her out of bed into the arms of someone who then carries her to the toilet. Behind her bed hangs a photo, famous in my family, of her standing next to my grandfather on a beach, both of them in bathing suits. They're young and ecstatic to be together on the shore: my grandfather bares his teeth; she smiles a crooked smile into the sun.

When my father visits, he takes her out for walks. Her right hand is strapped to a small platform attached to her wheelchair so it doesn't fall and hurt her. My father moisturizes her skin, which often becomes dry, applies sunscreen, and puts a baseball cap on her head. If it's chilly outside, he tucks blankets across her lap. He wheels her outside, where they follow a sidewalk around the facility's campus. On their walk, they pass two small ponds beside which they stop to do some bird-watching. They continue down a residential street, and as they come out to the main road, the tree cover disappears. My father adjusts my grandmother's cap.

Back inside, she begs him, as best she can, not to leave her. She tells him she wants to die. My father kisses her and leaves.

The summer after my grandmother's stroke, I decided to paint her nails. When I got to her room, I discovered the set of acrylics she'd been wearing at the time of her stroke still hadn't been removed. Between her overgrown cuticles and the tops of the fake nails was a gulf of keratin. I asked her if she minded me painting over them. After working for weeks, she'd managed to learn to huff, "Yes." We sat in the common area, which smelled of antiseptic and bananas, before a dark TV that reflected us in its glass. I had brought two bottles of polish—hot pink and electric blue. She selected the pink. I talked while I painted. "I got pink on your wheelchair, Grandma. Do you think they'll care?" Silence. "This remover stinks, huh?" Silence.

When I returned in the fall, I found she'd been moved to a new room. I followed a receptionist through back hallways of the building, through the kitchen to the employee elevator, which I took to the third floor. The new room was identical to the old one, but had a window that cast sun across my grandmother's bed. She was watching Turner Classic Movies beneath a white afghan that used to live on her family room sofa. She'd reached a point in her recovery where her right leg would move of its own volition, but when she tried to move it intentionally, it stayed put. The film on the TV told the story of a rodeo cowboy and his beautiful lover, worried for his life. It was nearing the end, and together we watched the scene of the cowboy's last ride. He mounts the bronco; they open the gate. His lover watches from the sidelines, her knuckles white, her eyes unblinking. The bronco bucks and the cowboy is thrown to the ground—the lover screams; she closes her eyes. In the last shot, the cowboy lies motionless in the dirt.

I know what my husband will look like when he's old. I know how the skin of his face will hang from his bones. His cheeks will hollow—he's already thin. The lines around his mouth will grow more defined. Folds will deepen across his forehead, worry lines. His hair will never turn fully gray or fall out—he's Italian. Tiny crinkles around his eyes will fan out like sun rays. His eyes will still be green and curious.

Three weeks ago, my husband was diagnosed with testicular cancer. A week later, a surgeon removed his testicle, and in one month, he will begin the first of four rounds of chemotherapy. His hair will fall out. Eating will be a challenge. We're asking questions we've never considered before: How will his history of heavy smoking affect his ability to receive chemotherapy since chemotherapy puts him at risk for pulmonary fibrosis? How long will he be sick after chemo ends? What are the chances we'll have to do it again? How will we pay rent in the meantime? How do we go about freezing sperm so that we can have children?

Sometimes I make him promise he won't die. I tell him I'll die, too, if he goes anywhere. Maybe this is cruel.

It used to upset my husband when I bothered him about smoking. I told him I was afraid of someday not hearing him breathe while he sleeps. I worried about how painful it would be for him to struggle to inhale, how it would likely be even more painful to exhale. I worried he'd suffer and that I'd have to watch him. That our future children would learn too soon of their own mortality, through their father's. Even before I know them, I want to shield them from this pain.

My parents have a pact that my husband and I adopted: that one of them will kill the other if necessary. If they're brain-dead. If they lose the powers of speech and movement. If they're going to be a burden on each other. If there's no possibility of a life of quality. If, going forward, life will only be suffering. An act of mercy.

Watching my father care for his mother, I wonder what I will do for him. What he will need when the unpredictable happens. How I will give it to him. How to prepare. I think about how little I have now, how I'm still so dependent on my parents for so many things—help with our taxes, marital wisdom. I will need to outgrow myself. I will transform in unthinkable ways. It will teach me the true meaning of dignity.

AFTER MY GRANDFATHER DIED, I IMMERSED MYSELF IN DEATH. I WORKED AT A BOOKSTORE AT the time and used my staff discount to buy books constantly: an anthology of essays about grief by famous writers, a popular science book about corpses. I read about the funeral industry and eco-friendly ways to dispose of bodies. I became fascinated by ghosts. I studied time, eternity, and the number zero. My grandfather's death was my first death. It moved me in a way I'd never felt before, and I thought that if someone, some authority outside of myself, could explain the feeling, then my experience of it would seem natural. But nothing got at the core of what I was feeling. Each experience of grief I encountered was different, and none resembled my own. Then I reread *The Velveteen Rabbit.*

My parents read me this book every night as a child. I think it was their favorite, too: I can still hear the sound of my mother reading it, her alto register, lilting and musical. Our abridged Golden Books edition still sits on the bookshelf in my apartment. Its cover is glossy with a colored pencil drawing of a stuffed rabbit sitting in a thicket of ferns and ivy with a sprig of blackberries over its shoulder. Inside, it tells the story of a boy who receives a Velveteen Rabbit for Christmas. At first, the Boy prefers other toys: mechanical ones, like the engine and the model boat. The Rabbit is shy and self-conscious; he knows that his velveteen fur and his sawdust filling are the stuff of simple toys. The other more glamorous toys make fun of him—all except for the Skin Horse, who has been in the nursery the longest and is therefore the wisest of all the toys. One day, when they're alone, the Skin Horse tells the Rabbit what it means to become Real. It's not how you're made, he says. It's a thing that happens to you when someone loves you enough. Oftentimes it hurts. But once you're Real, you're Real forever.

The Rabbit is fascinated by this idea of being Real. But he doubts it's likely to happen to him, being plain as he is. Then a toy goes missing at bedtime, and the Boy is given the Rabbit to sleep with. The two become inseparable. When the Boy falls ill with scarlet fever, the Rabbit stays by his side, day and night.

As a child, it seemed to me that the Boy's scarlet fever sat someplace outside the Rabbit's story, that the fever had little to do with the Rabbit. It was something that the Rabbit had to live through, put up with, because of the Boy. Now I understand that it wasn't for the Boy's sake that the Rabbit stayed with him through his fever; it was for the Rabbit's own sake, too. Afraid he will be removed from the boy's side, the Rabbit hides from sight under the bedclothes, perfectly still. As much as the Boy needs him, the Rabbit needs to be needed as well.

When the Boy is well again, the Rabbit is thrown in a sack with the rest of the contaminated nursery toys. *Of what use was it to be loved and lose one's beauty and become Real if it all ended like this?* he thinks. He sheds a tear, and where it falls, there grows a flower, and from it springs a Fairy. The Fairy takes him to the forest and kisses him on the nose. Then she sets him free to run wild with the rest of the rabbits.

My father has his father's eyes, as do I. That first Thanksgiving without my grandfather, I asked my father to read me *The Velveteen Rabbit*. We sat on the couch and I listened, slipping in and out of sleep, as his blue eyes moved across the pages with familiarity and he performed the characters in his gentle tenor. He stopped now and again to ask questions, amused at reading to me again as though I were still small. I thought of the rabbit my grandmother gave me. Over time, she accepted the shape of my body and gave up her own. As she wore down, I'd slip my finger through a hole in her seams and find a thread to repair the damage. Finally we reached a stage where I'd sit her on the bedside table at night instead of bringing her to bed, afraid I'd ruin her. Days went by without my lifting her. One day, she disappeared.

Now my rabbit lives in a box along with our winter coats. Each year when the weather turns cold, she comes tumbling out in a ball of familiar scent. I hug her close: my neck still knows the way her arms wrap around it; my right cheek knows the feeling of hers, soft against my own. I look into her neutral expression, which has turned downward over the years, and I think of all those nights drifting off in my childhood bed. Then I put her back in the box, out of sight, not forgotten.

Before: An Inventory

Written on the occasion of turning thirty.

June, my birthday.

Botanic Garden bees in the roses, white dog at the in-laws', roaches in my apartment—Brooklyn, goldfish at my parents' house—Largo, cats pissing in the laundry, lizards on the porch (looking weathered), jays in the roses, Sunken Gardens kookaburra, cockatoo, flamingo, stray dogs on the freeway off-ramp.

May.
Entering Florida.

Road trip.

Reverse west /
Grackles on the power lines—Austin, emu on the roadside—Marfa, vultures (white-tipped wings) in the headlights, bats on

the highway, lambs in truck beds, horses grazing in New Mexico, grazing milk cows in Tucson—and on and on the bluebells.

Up the coast /
Molting-season elephant seals, elk in the redwoods, sapphire jay at the general store, snails on the rocks in Seattle, alien mouths of barnacles, starfish, sea anemones, rumors of black whales on black rock beaches, sexually frustrated cat in Tacoma licking itself on the bookstore floor.

Reverse east /
Bison in Wind Cave (forest fire), ibex at Rushmore, gray Great Dane (named Zeus)—the Twin Cities, hawks over Iowa soy fields, Lab at the Hunts' house, beetles mating on the neck-high corn, farm with alpacas (long-haired and matted), white bunny kept to the rug—Chicago.

Flight /
Mexican cows crossing the roadway, rescued dogs in the host's house—San Miguel, strays in the brick door frames, donkeys hauling tequila to a wedding, lichen on the ruins, legend of rose quartz deep beneath the city—*Didn't we see this white dog yesterday?*

Home (New York) /
Cat's paws in my purse on the scent of—, white dogs (two now) at the in-laws', turkeys in the yard on Valentine's, January snowstorm and a black cat in Whitney's Brooklyn kitchen (dinner), one-eyed cat on a new friend's new roof (Brooklyn, New Year's).

Fall, driving.

Pennsylvania gas station (hunting collage), milk cows pissing in their drinking pond, flies on cows' noses, flies in their eyes, cats (coddled) in the Flores apartment, truck stop (barbed wire) horses, trucks of chickens, prairie dogs' calls (high and clear), broken sun rays through the mountains leaving Denver.

Fly on my cupcake—Silver Lake, cocker spaniel at a candlelit backyard party, cats on a pepper farm chewing fleas, wood-colored moths on the farmhouse, pit bulls licking our hands through fence slats, sexually frustrated cat in Los Feliz humping my husband's shoes, skinny palm trees reaching for more of *that*.

July Fourth.

Maltese sitting in Fort Greene (fireworks), mutt in the entrance sniffing our knees, bedbugs in the box springs, Botanic Garden dragonflies, bees in the roses, koi in the moss pond, inchworm in the super's flowers, colonies of blue wasps at the windows, sparrows on the sidewalk eating day-old bread left out by the neighbors.

June, my birthday.

White dogs on the carpet, wasps at my parents' windows, ants in the bathroom, maggots in the toothpicks, orb weavers in the orchids on the back porch (just refinished), crabs coming up (tiny ripples) from the mangrove mud, warnings of warm-water sharks at nighttime—*Don't go swimming naked in the dark*.

Infant at the ex's (Asheville), black Lab at the hidden river (Bat Cave), ticks at the waterfall, horses at the general store, Chihuahua in Emily's lap at the Brooklyn Flea, new white dog at the in-laws' keeping the old one company as it dies, wall of vines in my neighborhood hiding nests.

January,
a wanted but untimely arrival:
a new human animal.

Dying white dog at the in-laws',
Christmas.

Polo in the Hamptons—late summer, clams in the inlet, mussels in a net sack, snakes at the Pollock House—late-night, knee-high grass (rumors of leeches), mud in our cutoffs, ticks in our socks, belly-up under stars (stoned) on the concrete: *See that constellation? Andromeda the Insolent.*

Brooklyn Botanic Garden (birthday), bees in the roses, koi in the moss pond, bunny (Easter) at the bookstore (Shakespeare), kittens clawing on the Wileys' new carpet, Los Angeles hummingbird drinking sugar from the feeder, moving on to bougainvillea—*Where do flies go in January?*

Killing flies (worn copy of *On Being Blue*) in the kitchen—Brooklyn, goats at the bookstore story hour, grandfather's death (cancer, late summer)—robin chicks at the front door, deer stalking the yard like reapers, pulling at grass, awaiting his departure—deer in the headlights in Montauk (our first anniversary).

Fourth of July.

San Diego boardwalk vendor hocking parrots, hummingbirds in the cacti flowers—Sedona, sand fleas (half covered) in the mud—Florida, butterfly on my father's shirt at my graduation, bees on knotweed's purple blossoms, and an old mastiff (asleep) in an East Williamsburg storefront.

Winter (our first apartment).

Mouse in the linen, roaches in the garbage, flies on the ceiling, feet of our neighbors, new husband's old friends' old house cat, vegan beagle (named Lady), hog on a leash—Central Park, turkey at the in-laws', in-laws' dying green bird (its broken legs), wedding that summer in the Van Nuys courthouse.

Sexually frustrated cat in Los Feliz humping my lover's shoes, lone cormorant (contemplating waves) on the Malibu beach (magic hour), coquinas in the sunset sand, beer on the Venice boardwalk (our vows in a bottle), Mojave Desert—pissing on cacti, black and white sheep on the road to California. *Where do the birds go when it blizzards?*

White dog (deaf) at my lover's house, dogs at the bookstore, (walking, a new city) Chinatown fish in milky windows,

driving the U-Haul trailer up I-95 with my boyfriend
(for the moment) (truck stop fuck, mosquitoes),
sick duck on the day we're leaving:
Call the seabird sanctuary.

Bunny in a hoarder's house, rescued red-eared slider in the street, spiders at the kids' museum, opossum on the road stunned and blinking, puppy in a cardboard box in Ginnie Springs, snakes in a bucket, pelicans at the pier, fishermen throwing guts up to the gulls, pug at the lover's house, mix at the boyfriend's mother's house, rat in the classroom, and a hawk in the school yard with the children.

Worms in the bus circle, bees in the class garden, butterflies at the art museum (field trip), Florida Aquarium—rays in the touch tank, anemones, seahorses, state fair monkey posing for photos, goldfish in a sandwich bag, husk of a horseshoe crab, cat in my one-room apartment, termites' wings on my face upon waking (that give in the floor, the smell of compost), and the next-door beagle—*Lonelier than his owner.*

Alligator with her young in her mouth, bird railing against the window with blood on its beak.

New school year.

Orange cat at my parents' house replacing the dead one,
striped cat I abandoned to live with my parents,
abandoned (overdose) pit bull next-door,
chickens at my boyfriend's parents',
black runt cat beneath the porch,
giraffe (onyx-eyed) at the zoo.

Return to Florida.

Mixed-breed (road dog) in a hobo camp—the Allegheny River,
stray cats in a squat house (shit on the floor)—Pittsburgh,
spiders in the sliding glass doors of a stranger's home,
(follow trails of fairy houses toward the cliffs),
mangled-tailed cat in Maine for the summer,
boxer puppies (sleepy-eyed) in a closet.

Run away.

Boyfriend's black-and-white mix—confined to a crate, poodle on Atlantic Beach (cramming for finals), owl on my parents' back porch (home for summer), strays in Athens, crabs on Crete, lover's black-and-white mix (and his girlfriend's Chihuahua), two orange cats sniffing the fish, Kelly running horses at the stables, counting rats in the New York subway (my first time).

Snails on the neighbor's dock that final summer, Asia's sugar glider in the sheets (her first apartment), Miles's iguanas in tanks (by the bongs), looming campsite banana spiders, Bonnie's black Lab's fur on the couch at Thanksgiving, gnats in the wet heat, swans on the restaurant pond on Sundays, cats at the SPCA (community service), saving kittens from a crack house (skipping seventh period).

Sugar glider down a shirt in the school cafeteria, crickets in the Petland sold to feed the lizards, fish, parrots, bunnies, turtle from a roadside vendor, goldfish in a plastic bag—Florida State Fair, pit bull (heard us fucking) at Danny's mother's, asthmatic cat at the weed dealer's, Asia's six cats eating organic chicken (she prepares), roaches in the kitchen of the coffee shop after school.

Trail ride through Yosemite mountains, herons on Alcatraz, puffins off the coast, yellow Lab loose again from the neighbor's, chocolate Lab loose again in our crotons, mutts dragging leaves through the back door (Stephanie's), naked conch sucking our wet palms (slime on our fingers), freshman science class slicing earthworms, leathery backs of guitarfish circling the tank—Florida Aquarium.

Milking cows on a farm—Ohio, crushed lizard (regret) beneath my bike tire, lizard beneath my palm on the back porch, tree frog at my father's office, ants (making houses) on the sidewalk, rat rotting in the leaky ceiling, black-and-white blind guinea pig at Amanda's house, ducks in the backyard, frozen cats in the science lab drawers, pinning open earthworm hearts, dead lizards (dried up) in our closets.

Barnacles on the neighbor's dock,
parrot on the pool deck (next door),
new orange cat (replacing the lost one),
geese in the cemetery pond (where I'm alone),
spiders on the river-rock porch of our new home.

New school, new friend's
bunny hutches, tank of seahorses,
gooseberry jellies (don't touch) on the beach,
cat in the Christmas tree, gray-and-white cat at Dené's,
classroom guinea pig, collie, and a bichon show in Cleveland.

Chow (cracked tooth) at my neighbor's house, koi in the churchyard, puppy (rescued) in a backyard pool, fish egg clusters at the babysitter's, four angry dogs behind chain-link, broken monarch wing on the playground at recess, skinny black garden snake stiff on the sidewalk, new white-and-tan hamster (replacing the dead one) in a cage in the garage, playing fetch with a new tabby kitten named Skittles.

Iguanas at the sitter's (keep dying), Polaroid of a half dog (back half) (alligator) in the backyard, white Sunday school hamster with red eyes, elephant at the Renaissance fair, lizard dropped into the AC unit (that noise) on the side of the house, lizard with three tails in Katie's backyard on Britton Street, six basset hounds barking in the yard behind, five cats slinking around in the garage.

Lizard eggs in the babysitter's back fence, toads hiding in a live oak in the school yard, wasp (finger stung) in the neighbor's sunflower, alligator in a retention ditch between houses, bags of mosquito larvae passed off as minnows, bags of potato chips emptied onto the front lawn (attracting gulls), bees in the monkey bars after school, carpenter ants in a vial in the fridge, raccoon babies in the hollow tree in the backyard.

Goats at the petting zoo eating carrots, fiddler crabs in the mangrove tangle (languid summers), white Persian cat in a cardboard box (coming home from school), mossy rocks with earthworms slithering underneath, lizard in a jewelry box with a twirling ballerina, yellow jackets at the swing set (ice cream), wasps in a muddy nest on the front porch, litter of white mice at the house on Britton Street.

Chipmunks in my grandma's geraniums, bunny in the bushes darting away, grubs buried deep in the wet earth of the school yard, silkworms dangling from the mulberry tree, black-and-white caterpillar on a finger, brown dog by the fence at the edge of the playground, sand dollars at the beach and the stingray shuffle, cockatiel named Buddy at my father's office.

Flying roach in my bathroom (pink),
ducklings at the sitter's (Genevieve, ancient),
falling asleep on the rug to *MacGyver*, the taste of licorice,
and the lizard she tossed out into water just to watch it wriggle.

Pony at the preschool (picture day), chiggers in the moss, soapberry bugs on the rain tree, chickens in a yard. Cats at the house on Britton Street, finches in a cage, egrets on the lawn eating hot dogs, blind pug at Thanksgiving.

Slugs (dried up) on the sidewalk, snails in the grass, birds in a nest in the rain tree, white Persian cat. Cats in a house on Maple Street, skunk in a garage, little brown dog at a pool party, fat locusts in the yard.

Fireflies in Cleveland in the summer, mosquitoes in the spring, black-and-white dog at the neighbor's house, a feral cat (our first, Phoenix).

ACKNOWLEDGMENTS

Thank you first to my husband, David Formentin, for his infinite support and patience, and to my family for the same. To Erin Wicks and Adriann Ranta, for without them this book would not even be a concept. Rachel Hurn and Amy Gall, my trusted first readers. Cal Morgan, for taking a chance on me. To all who trusted me with their stories, and to all who helped me research and write them.

BIBLIOGRAPHY

PUBLISHED MATERIAL

Abkowitz, Alyssa. "A Resort with Less Glitz and No Kitsch." *Wall Street Journal*, February 14, 2013. http://www.wsj.com/articles/SB100014241 27887323696404578298124291819096.

Amway. "Amway Approved Provider—Yager Group." Accessed June 9, 2016. http://www.amway.com/about-amway/yagergroup.

———. *Business Reference Guide*. Ada, MI: Amway, 2015.

"Amway Function." YouTube video, 1:44, posted by trapparker4, July 11, 2011. https://youtu.be/lzR_bNK-w1o.

Associated Press. "Florida Pelicans Retire in Texas." *Ocala Star-Banner*, December 17, 1976.

Bayou Club. "Welcome to the Bayou Club." Accessed May 30, 2016. http://www.bayouclubgolf.com/.

Bennett, Laurie. "The Ultra-Rich, Ultra-Conservative DeVos Family." *Forbes*, December 26, 2011. http://www.forbes.com/sites/lauriebennett/2011/12/26/the-ultra-rich-ultra-conservative-devos-family/#4e2ae3af2c86.

Berman, Dennis K. "Inside the Amway Sales Machine." *Wall Street Journal*, February 15, 2012. http://www.wsj.com/articles/SB1000142405297020 4062704577223302734609434.

Boca Beacon. "Beatrice Busch Promotes Sanctuary." April 1, 1982.

Bothwell, Dick. "Seabird Sanctuary Gains Recognition—but the Bills!" *St. Petersburg Times*, August 2, 1974.

Braden, Charles. *Spirits in Rebellion: The Rise and Development of New Thought*. Dallas: Southern Methodist University Press, 1987.

Bradsher, Keith. "And One Who Tries to Work Things Out." *New York Times*, October 31, 1995. http://www.nytimes.com/1995/10/31/us/and-one-who-tries-to-work-things-out.html.

Burstein, Rachel, and Kerry Lauerman. "She Did It Amway." *Mother Jones*, September/October 1996. http://www.motherjones.com/politics/1996/09/she-did-it-amway/.

Butterfield, Stephen. *Amway: The Cult of Free Enterprise*. Boston: South End Press, 1985.

Cady, H. Emilie. *Lessons in Truth*. Unity Village, MO: Unity Books, 2007.

Cappiello, Dina. "Brown Pelicans off the Endangered Species List." *San Francisco Gate*, November 12, 2009. http://www.sfgate.com/green/article/Brown-pelicans-off-endangered-species-list-3211230.php.

Champion Turf Farms. "Champion: The Ultimate Bermudagrass Putting Surface." Accessed May 30, 2016. http://www.championturffarms.com/champion/.

Chapman, Michael W. "Maya Angelou: 'God Loves Me'—'That's Why I Am Who I Am.'" CNSNews.com, May 28, 2014. http://www.cnsnews.com/news/article/michael-w-chapman/maya-angelou-god-loves-me-s-why-i-am-who-i-am.

Charity Navigator. "Suncoast Seabird Sanctuary: Historical Ratings." Accessed November 11, 2015. https://www.charitynavigator.org/index.cfm?bay=search.history&orgid=5141.

Coffey, Brendan, and David de Jong. "Amway Billionaires Debut After Bad Year for Direct Seller." *Bloomberg*, February 10, 2015. http://www.bloomberg.com/news/articles/2015–02–10/amway-billionaires-debut-after-bad-year-for-direct-seller.

Corporation Wiki. "12388 Starkey Road, LLC." Accessed October 29, 2015. https://www.corporationwiki.com/p/2fyqa1/12388-starkey-road-llc.

Crawford, Craig. "What About NBA's Homophobic Owner?" *Huffington Post*, last updated June 30, 2014. http://www.huffingtonpost.com/craig-crawford/what-about-nbas-homophobe_b_5236780.html.

Cross, Wilbur. *Amway: The True Story of the Company That Transformed the Lives of Millions*. New York: Berkley, 1999.

———. *Choices with Clout: How to Make Things Happen by Making the Right Decisions Every Day of Your Life*. New York: Berkley, 1995.

DeChant, Dell. "Myrtle Fillmore and Her Daughters: An Observation and Analysis of the Role of Women in Unity." In *Women's Leadership in Marginal Religions*, edited by Catherine Wessinger. Urbana: University of Illinois Press, 1993.

———. *Unity and History: Course One*. Clearwater, FL: Unity Progressive Press, 2012.

Delaney, Arthur. "How a Traveling Consultant Helps America Hide the Homeless." *Huffington Post*, March 9, 2015. http://www.huffingtonpost.com/2015/03/09/robert-marbut_n_6738948.html.

DeVos, Rich. *Compassionate Capitalism*. New York: Plume, 1994.

———. *Simply Rich: Life Lessons from the Cofounder of Amway*. New York: Howard Books, 2014.

Dewar's. "Dewar's Profiles: Ralph Heath." Magazine advertisement. Leo Burnett USA. Directed by Gene Kolkey, 1974.

Dick and Betsy DeVos Family Foundation. "About." Accessed May 30, 2016. http://www.dbdvfoundation.org/about/.

Douglas, Mark. "Embattled Suncoast Seabird Sanctuary Founder Pleads Not Guilty." News Channel 8, May 26, 2016. http://wfla.com/2016/05/26/embattled-suncoast-seabird-sanctuary-founder-pleads-not-guilty/.

———. "Foreclosure Filed over Seabird Sanctuary Warehouse." News Channel 8, January 24, 2013. http://www.tbo.com/pinellas-county/foreclosure-filed-over-seabird-sanctuary-warehouse-615693.

———. "Photos Show 'Deplorable' Conditions for Wildlife Held Captive by Suncoast Seabird Sanctuary Founder." News Channel 8, last updated May 24, 2016. http://wfla.com/2016/05/23/photos-show-deplorable-conditions-for-wildlife-held-captive-by-suncoast-seabird-sanctuary-founder/.

———. "Seabird Sanctuary Workers Say Donations Disappeared." News Channel 8, November 2, 2012. http://www.tbo.com/news/suncoast-seabird-sanctuary-workers-say-donations-disappeared-552619.

———. "Suncoast Seabird Sanctuary Finances Face Review." News Channel 8, October 14, 2012. http://www.tbo.com/news/florida/suncoast-seabird-sanctuarys-finances-face-review-534007.

———. "Suncoast Seabird Sanctuary Must Pay $21,336 in Back Wages." News Channel 8, November 15, 2012. http://www.tbo.com/pinellas-county/suncoast-seabird-sanctuary-must-pay—in-back-wages-565553.

East, Jon. "'Nature Lovers' Plan Petition Drive for Bird Sanctuary Land Sale." *St. Petersburg Evening Independent*, October 13, 1982.

———. "Sod Farm Land to Be Considered for Landfill Use." *St. Petersburg Evening Independent*, June 27, 1984.

Easy Street. DVD. St. Petersburg, FL: Wideyed Films, 2006.

Eclectablog. "DeVos Family Responsible for HALF of Campaign Contributions to Michigan House Republicans in Last Quarter of 2015." February 2, 2016. http://www.eclectablog.com/2016/02/devos-family-responsible-for-half-of-campaign-contributions-to-michigan-house-republicans-in-last-quarter-of-2015.html.

Evertz, Mary. "300-Acre Estate Is Setting for Busch-Heath Wedding." *St. Petersburg Times*, August 6, 1982.

Florida Department of Children and Families. *Council on Homelessness 2015 Annual Report*. Tallahassee, FL: Council on Homelessness, 2015.

Forbes. "America's Top 50 Givers." Accessed October 5, 2016. www.forbes.com/top-givers.

———. "The World's Billionaires." Accessed October 5, 2016. www.forbes.com/billionaires.

Flying Free 2. DVD. Directed by Jerry Alan. Valrico, FL: Matheney Productions, LLC, 2012.

Foscarinis, Maria, Kelly Cunningham-Bowers, and Kristen E. Brown. "Out of Sight—Out of Mind?: The Continuing Trend Toward the Criminalization of Homelessness." *Georgetown Journal on Poverty Law & Policy* 6, no. 2 (1999): 145–164.

Frago, Charlie. "Road to Nowhere: Homeless Bused out of St. Pete, but Then What?" *Tampa Bay Times*, July 7, 2015. http://www.tampabay.com/news/humaninterest/road-to-nowhere-homeless-bused-out-of-st-pete-but-then-what/2236491.

Freeman, James Dillet. *The Story of Unity*. Unity Village, MO: Unity Books, 2007.

Gallagher, Peter B. "The Rise and Fall of Chief Jim Billie." *Gulfshore Life*, September 2010. http://www.gulfshorelife.com/September-2010/The-Rise-and-Fall-of-Chief-Jim-Billie/.

Gibeaut, John. "Domestic Violence Arrests May Slow Vicious Cycle." *Sarasota Herald-Tribune*, August 11, 1986.

Girardi, Steven. "Groups Work to Provide Permanent Housing for St. Pete's Homeless." *St. Petersburg Tribune*, August 16, 2015. http://www.tbo.com/pinellas-county/groups-work-to-provide-permanent-housing-for-st-petes-homeless-20150816/.

———. "'Tiny Houses' Help St. Pete Tackle Challenge to House Homeless Veterans." *St. Petersburg Tribune*, December 12, 2015. http://www.tbo.com/pinellas-county/tiny-houses-help-st-pete-tackle-challenge-to-house-homeless-veterans-20151212/.

Goff, Brian. "Injured Heron Finds Its Way Back to Seabird Sanctuary." TBNweekly.com, January 6, 2016. http://www.tbnweekly.com/editorial/outdoors/content_articles/010616_out-01.txt.

———. "New Bird Rescue Operation Fills a Void." TBNweekly.com, March 14, 2013. http://www.tbnweekly.com/pinellas_county/content_articles/031413_pco-01.txt.

Grand Rapids Press Staff. "Lawyers Say Their $20 Million Payment Is Fair for $100 Million Settlement in Amway Pyramid Scheme Lawsuit." Mlive.com, November 4, 2010. http://www.mlive.com/business/west-michigan/index.ssf/2010/11/lawyers_say_their_20_million_p.html.

Greene, Jackie. "In a Family Way." *St. Petersburg Evening Independent*, April 30, 1974.

Griffin, Bob. "Andrew's Island—The Third One." *Madeira Beach Communicator*, May 2012.

Hall, Katelyn. "Anheuser-Busch Brothers Tap into Dating App Market with Support from Mark Cuban." *Dallas Morning News*, February 14, 2015. http://www.dallasnews.com/business/small-business/20150214-anheuser-busch-brothers-tap-into-dating-app-market-with-support-from-mark-cuban.ece.

Hallifax, Jackie. "Justices Rap Domestic Violence Law." *Ocala Star-Banner*, October 27, 1994.

Harley, Gayle. *Emma Curtis Hopkins: Forgotten Founder of New Thought*. Syracuse, NY: Syracuse University Press, 2002.

Heben, Andrew. *Tent City Urbanism: From Self-Organized Camps to Tiny House Villages*. Eugene, OR: Village Collaborative, 2014.

Helderop, Brandon. "Detroit Can Learn a Thing or Two from Grand Rapids." *Huffington Post*, April 1, 2013. http://www.huffingtonpost.com/brandon-helderop/detroit-can-learn-from-grand-rapids_b_2981627.html.

Holan, Mark. "Pasco County 400-Acre Seabird Sanctuary for Sale." *Sarasota Herald-Tribune*, December 7, 2002.

Karas, Tania. "At 100, Helen Heath Puts Her Life into the Suncoast Seabird Sanctuary." *Tampa Bay Times*, January 12, 2010. http://www.tampabay.com/news/humaninterest/at-100-helen-heath-puts-her-life-into-suncoast-seabird-sanctuary/1064896.

Keyes, Scott. "Courts Are Striking Back Against the Criminalization of Homelessness." *Think Progress*, October 22, 2015. http://thinkprogress.org/economy/2015/10/08/3709492/homeless-panhandling-bans-unconstitutional/.

———. "Criminalizing Homelessness Can Now Cost Cities Federal Money." *Think Progress*, September 22, 2015. http://thinkprogress.org/economy/2015/09/22/3704274/hud-homelessness-criminalization-funding/.

Knape, Chris. "Amway Agrees to Pay $56 Million, Settle Case Alleging It Operates a 'Pyramid Scheme.'" Mlive.com, November 3, 2010. http://www.mlive.com/business/west-michigan/index.ssf/2010/11/amway_agrees_to_pay_56_million.html.

Kneale, Klaus, and Emily Lambert. "Climb to the Top." *Forbes*, July 24, 2008. http://www.forbes.com/forbes/2008/0811/050.html.

Kroll, Andy. "Meet the New Kochs: The DeVos Clan's Plan to Defund the Left." *Mother Jones*, January/February 2014. http://www.motherjones.com/politics/2014/01/devos-michigan-labor-politics-gop.

Larimer, Mary E., et al. "Health Care and Public Service Use and Costs Before and After Provision of Housing for Chronically Homeless Persons With Severe Alcohol Problems." *Journal of the American Medical Association* 301, no. 13 (April 1, 2009): 1349–1357.

Lemkowitz, Florence. "A Sanctuary for Seabirds in Florida." *New York Times*, May 25, 1980.

Lindberg, Anne. "Foreclosure Suit Filed against Suncoast Seabird Sanctuary Property." *Tampa Bay Times*, January 24, 2014. http://www.tampabay.com/news/humaninterest/foreclosure-suit-filed-against-suncoast-seabird-sanctuary-property/1272116.

———. "Mistake Gives Suncoast Seabird Sanctuary Owner Years of Tax Breaks." *Tampa Bay Times*, October 26, 2012. http://www.tampabay.com/news/humaninterest/mistake-gives-suncoast-seabird-sanctuary-owner-years-of-tax-breaks/1258508.

———. "Seabird Sanctuary Founder Sued Twice, Accused of Selling Car to Two Different Buyers." *Tampa Bay Times*, October 29, 2014. http://www.tampabay.com/news/humaninterest/seabird-sanctuary-founder-sued-twice-accused-of-selling-car-to-two/2204270.

———. "State Files Tax Lien against Troubled Suncoast Seabird Sanctuary." *Tampa Bay Times*, February 13, 2013. http://www.tampabay.com/news/humaninterest/state-files-tax-lien-against-troubled-suncoast-seabird-sanctuary/1275107.

———. "Suncoast Seabird Sanctuary Founder Charged with Workers' Compensation Fraud." *Tampa Bay Times*, November 5, 2013. http://www.tampabay.com/news/humaninterest/suncoast-seabird-sanctuary-founder-charged-with-workers-compensation-fraud/2150939.

———. "Suncoast Seabird Sanctuary No Longer Taking in Injured Birds." *Tampa Bay Times*, January 30, 2013. http://www.tampabay.com/news/

humaninterest/suncoast-seabird-sanctuary-no-longer-taking-in-injured-birds/1273020.

———. "Suncoast Seabird Sanctuary Warehouse to Be Auctioned to Pay Debt." *Tampa Bay Times*, December 13, 2013. http://www.tampabay.com/news/localgovernment/suncoast-seabird-sanctuary-warehouse-to-be-auctioned-to-pay-debt/2156940.

———. "U.S. Labor Department Investigating Suncoast Seabird Sanctuary." *Tampa Bay Times*, October 12, 2012. http://www.tampabay.com/news/humaninterest/us-labor-department-investigating-suncoast-seabird-sanctuary/1256197.

Madison, Michael. "Neighbors Voice Trouble with Pinellas Safe Harbor Shelter." *Clearwater Patch*, July 15, 2011. http://patch.com/florida/clearwater/neighbors-voice-troubles-with-pinellas-safe-harbor-shelter-2.

Marbut, Robert. "Seven Guiding Principles." Marbut Consulting, October 23, 2010. http://www.marbutconsulting.com/Seven_Guiding_Principles_FQ.html.

Martinez, Shandra. "$1.2 Billion in Donations Puts DeVos Family in Forbes Top Philanthropy List." Mlive.com, October 1, 2015. http://www.mlive.com/business/west-michigan/index.ssf/2015/10/13b_in_donations_puts_devos_fa.html.

———. "Amway's 2015 Revenues Fall to Lowest Level in 5 Years." Mlive.com, February 3, 2016. http://www.mlive.com/business/west-michigan/index.ssf/2016/02/amways_2015_revenues_fall_to_l.html.

———. "How and Why the DeVos Family Gives Away Billions." Mlive.com, January 4, 2016. http://www.mlive.com/business/west-michigan/index.ssf/2016/01/devos_family_donations.html.

Mary Baker Eddy Library. "The Life of Mary Baker Eddy." Accessed July 13, 2016. http://www.marybakereddylibrary.org/mary-baker-eddy/the-life-of-mary-baker-eddy/.

Matthews, Downs. "Volunteers Rescue Injured Wildfowl." *Smithsonian*, August 1974.

McMahon, Patrick. "Rare Brown Pelican Born at Suncoast Sanctuary." *St. Petersburg Times*, May 5, 1975.

Meacham, Andrew. "Authorities Issue 59 Violations against Suncoast Seabird Sanctuary over Animal Care." *Tampa Bay Times*, May 6, 2014. http://www.tampabay.com/news/humaninterest/authorities-issue-59-violations-against-suncoast-seabird-sanctuary-over/2178592.

Meinhardt, Jane. "'Double-Cross' Allegations over Site for Sanctuary." *St. Petersburg Evening Independent*, September 28, 1982.

Melton, J. Gordon. "Emma Curtis Hopkins: A Feminist of the 1880s and Mother of New Thought." In *Women's Leadership in Marginal Religions*, edited by Catherine Wessinger. Urbana: University of Illinois Press, 1993.

Michaelson, Jay. "The $1-Billion-a-Year Right-Wing Conspiracy You Haven't Heard Of." *Daily Beast*, September 25, 2014. http://www.thedailybeast.com/articles/2014/09/25/the-1-billion-a-year-right-wing-conspiracy-you-haven-t-heard-of.html.

Miller, Betty Jean. "Beatrice Busch-Keefe." *St. Petersburg Evening Independent*, March 13, 1982.

———. "Pelican Venture." *St. Petersburg Evening Independent*, December 5, 1978.

Missio Dei. "What Is the Missio Dei?" Accessed July 11, 2015. http://themissiodei.com/.

Mitchell, Robin. "Good Birds Make Good Neighbors." *St. Petersburg Evening Independent*, October 12, 1978.

Mlive.com. "Where DeVos Family Donates Millions." January 4, 2016. http://www.mlive.com/business/west-michigan/index.ssf/2016/01/where_rich_devos_and_his_famil.html#0.

Morrill, Jim, and Nancy Stancill. "Amway the Yager Way." *Charlotte Observer*, March 19, 1995.

Morrow, Alison. "Changes Expected in June Will Address Issues Plaguing Pinellas Safe Harbor Homeless Shelter." ABC Action News, WFTS Tampa Bay, May 13, 2013. http://www.abcactionnews.com/news/region-pinellas/changes-expected-in-june-will-address-issues-plaguing-pinellas-safe-harbor-homeless-shelter.

Mullane Estrada, Sheila. "Financial Problems Strain Suncoast Seabird Sanctuary." *Tampa Bay Times*, February 16, 2010. http://www.tampabay.com/news/localgovernment/financial-problems-strain-suncoast-seabird-sanctuary/1073656.

———. "Suncoast Seabird Sanctuary on Indian Shores Gets $100,000 Surprise from Qatar." *Tampa Bay Times*, May 1, 2012. http://www.tampabay.com/news/humaninterest/suncoast-seabird-sanctuary-in-indian-shores-gets-100000-surprise-from-qatar/1227739.

National Alliance to End Homelessness. *The State of Homelessness in America 2015: An Examination of Trends in Homelessness, Homeless*

Assistance, and At-risk Populations at the National and State Levels. Washington, DC: 2015.

National Coalition for the Homeless. *Substance Abuse and Homelessness.* Washington, DC: 2009.

National Law Center on Homelessness and Poverty. *Cruel, Inhuman, and Degrading: Homelessness in the United States under the International Covenant on Civil and Political Rights.* Washington, DC: 2013.

———. *No Safe Place: The Criminalization of Homelessness in U.S. Cities.* Washington, DC: 2014.

———. *"Simply Unacceptable": Homelessness and the Human Right to Housing in the United States.* Washington, DC: 2011.

National Law Center on Homelessness and Poverty and National Coalition for the Homeless. *Homes Not Handcuffs: The Criminalization of Homelessness in U.S. Cities.* Washington, DC: National Coalition for the Homeless, 2009.

Neal, Terry. "Agencies Rated Poorly on Response to Abuse Cases." *Boca Raton News*, February 4, 1994.

New World Encyclopedia. s.v. "Mary Baker Eddy." Last modified September 12, 2013. http://www.newworldencyclopedia.org/entry/Mary_Baker_Eddy.

New York Times. "Health Guide: Hypersensitivity Pneumonitis." Accessed June 4, 2016. http://www.nytimes.com/health/guides/disease/hypersensitivity-pneumonitis/overview.html.

———. "Health Guide: Meningitis-Cryptococcal." Accessed June 4, 2016. http://www.nytimes.com/health/guides/disease/meningitis-cryptococcal/overview.html.

Newman, Rev. Lux. "The Radiant I Am by Emma Curtis Hopkins (Animated)." YouTube video, 25:27, posted by luxnewman, January 26, 2011. https://youtu.be/KHMKgMhCIRg.

Nohlgren, Steven, and Kevin DeCamp. "Tensions Brew over Plans for Homeless Shelter at the Pinellas Jail Complex." *Tampa Bay Times*, December 7, 2010. http://www.tampabay.com/news/localgovernment/tensions-brew-over-plans-for-homeless-shelter-at-the-pinellas-jail-complex/1138691.

Nordqvist, Christian. "Birds and Their Droppings Can Carry over 60 Diseases." *Medical News Today*, September 22, 2014. http://www.medicalnewstoday.com/releases/61646.php.

Obama, Michelle. "The First Lady Announces the Mayors Challenge to End Veteran Homelessness." YouTube video, 35:28, posted by the White House, June 4, 2014. https://youtu.be/8oJnYi6XPVU.

O'Donnell, Jayne. "Multilevel Marketing or 'Pyramid?' Sales People Find It Hard to Earn Much." *USA Today*, February 10, 2011. http://usatoday30.usatoday.com/money/industries/retail/2011–02–07-multilevelmarketing03_CV_N.htm.

Office of Community Planning and Development, US Department of Housing and Urban Development. *The 2015 Annual Homeless Assessment Report (AHAR) to Congress*. Washington, DC: US Department of Housing and Urban Development, 2015.

Orcutt, Ben. "Down on the Farm: Tours Aim to Provide Missing Piece to Story." *Northern Virginia Daily*, May 6, 2011. http://www.wvgazettemail.com/ap/ApLife/201105100104.

Peppard, Alan. "Dallas Bar the Eberhard from Brothers with Beer in the Family." *Guide Live*, July 8, 2015. http://www.guidelive.com/bars-and-cocktails/2015/07/08/opening-eberhard-latest-brothers-von-gontard.

Phillips, Nancy. "Spouse Abuse Laws Helping Victims." *Sarasota Herald-Tribune*, October 15, 1984.

Pinellas County Homeless Leadership Board. *2011 Point in Time (PIT) Count of Homeless Individuals in Pinellas County*. St. Petersburg, FL: 2011.

Pinellas County Homeless Leadership Board. "Mission Statement." Accessed July 11, 2015. http://www.pinellashomeless.org/About-the-HLB/Mission-History.

Pyke, Alan. "Local Officials Have Pushed to Criminalize Homelessness for Years. The Feds Are Starting to Push Back." *Think Progress*, August 18, 2015. http://thinkprogress.org/economy/2015/08/18/3692251/homelessness-criminalization-doj-usich-hud/.

Pyramid Scheme Alert. "Study of Ten Major MLMs and Amway/Quixtar." April 12, 2010. http://pyramidschemealert.org/study-of-ten-major-mlms-and-amwayquixtar/.

Raghunathan, Abhi. "Public Defender Will Stop Working Homeless Cases in St. Petersburg." *St. Petersburg Times*, January 31, 2007. http://www.sptimes.com/2007/01/31/Southpinellas/Public_defender_will_.shtml.

Reens, David. "DeVos Family Goes All-In on Marco Rubio GOP Presidential Race." Mlive.com, February 25, 2016. http://www.mlive.com/news/grand-rapids/index.ssf/2016/02/devos_family_goes_all-in_on_ma.html.

Reynolds, Matt. "Amway's Claims Are Malarkey, Distributors Say." *Courthouse News*, April 29, 2014. http://www.courthousenews.com/2014/04/29/67407.htm.

Rogers, Alan. "Mary Baker Eddy and the American Dream." *We're History*, November 4, 2014. http://werehistory.org/mary-baker-eddy/.

Roth, Matt. "Dreams Incorporated." *Baffler* 10 (1997). http://thebaffler.com/salvos/dreams-incorporated.

San Diego, Bayani, Jr. "Ali's Wonderland a Labor of Love." Inquirer.net, June 16, 2011. http://entertainment.inquirer.net/3428/ali%E2%80%99s-wonderland-a-labor-of-love.

Santa Lucia, Ray. *Pinellas County Point in Time Homeless Report: 2015.* St. Petersburg, FL: Pinellas County Homeless Leadership Board, 2015.

Satter, Beryl. *Each Mind a Kingdom: American Women, Sexual Purity, and the New Thought Movement, 1875–1920.* Berkeley: University of California Press, 1999.

St. Louis Magazine. "All the Kings' Children." September 2015.

St. Petersburg Evening Independent. "Council Delays Gateway." February 8, 1974.

St. Petersburg Times. "Birds: Our Responsibility." December 23, 1974.

———. "Council Says It's Willing to Meet Nov. 10 with Seabird Sanctuary Founder." November 2, 1982.

Stanley, Kameel. "Officials Once Again Mulling over Homeless Problem in St. Pete, Pinellas County." *Tampa Bay Times*, July 14, 2014. http://www.tampabay.com/news/localgovernment/officials-once-again-mulling-over-homeless-problem-in-st-pete-pinellas/2187247.

———. "Homeless Issues Again Becoming Problem in St. Petersburg, Consultant Says." *Tampa Bay Times*, June 5, 2014. http://www.tampabay.com/news/localgovernment/homeless-issues-again-becoming-a-problem-in-st-petersburg/2183195.

Stewart, Nikita. "Obama Will Seek $11 Billion for Homeless Families." *New York Times*, February 6, 2016. http://www.nytimes.com/2016/02/09/nyregion/obama-to-propose-11-billion-to-combat-family-homelessness.html?_r=0.

Stiff, Robert. "City's Stand on Sanctuary Is for the Birds." *St. Petersburg Evening Independent*, December 1, 1982.

Stroud, Matt. "The Indian Express: How an Amway CEO Landed behind Bars Halfway around the World." *Verge*, January 18, 2013. http://www.theverge.com/2013/6/28/4472608/the-indian-express-pyramid-scheme-investigations-amway-herbalife.

Suncoast Seabird Sanctuary. *Fly Free*, newsletter, edited by Tara L. Gallagher, Summer 2008.

———. *Fly Free*, newsletter, edited by Jennie Hale, Spring/Summer 2014.

———. "Our Birds . . . By the Numbers." February 8, 2012. Accessed via the Internet Archive Wayback Machine on September 21, 2015. https://web.archive.org/web/20120208090710/http://www.seabirdsanctuary.com/Home_Page.html.

———. "Volunteers." October 24, 2008. Accessed via the Internet Archive Wayback Machine on June 3, 2016. https://web.archive.org/web/20081024051951/http://www.seabirdsanctuary.com/Volunteers.html.

Tampa Bay Times. "Memorial for Suncoast Seabird Sanctuary Employee Set for Sunday." September 8, 2010. http://www.tampabay.com/news/humaninterest/memorial-for-suncoast-seabird-sanctuary-employee-set-for-sunday/1120338.

Taylor, Mel. "Save Suncoast Seabird Sanctuary 2012." YouTube video, 12:04, posted by Get Smart Digital, June 19, 2012. https://youtu.be/fOJAzoN_OqQ.

Thomas, Robert McG., Jr. "August A. Busch, Jr. Dies at 90; Built Largest Brewing Company." *New York Times*, September 30, 1989. http://www.nytimes.com/learning/general/onthisday/bday/0328.html.

Thompson, Stephen. "Pinellas Authorities Evaluating Homeless Diversion Program." *St. Petersburg Tribune*, October 28, 2013. http://www.tbo.com/pinellas-county/pinellas-authorities-evaluating-homeless-diversion-program-20131028/.

———. "Pinellas Launching Program to Deal with Chronically Homeless." *St. Petersburg Tribune*, July 8, 2013. http://www.tbo.com/pinellas-county/pinellas-launching-program-to-deal-with-the-chronically-homeless-20130708/.

Ulferts, Alisa, and Abhi Raghunathan. "Police Slash Open Tents to Roust the Homeless." *St. Petersburg Times*, January 20, 2007. http://www.sptimes.com/2007/01/20/Southpinellas/Police_slash_open_ten.shtml.

Unofficial Amway Wiki. "Yager, Dexter & Birdie." Last modified January 5, 2015. http://www.amwaywiki.com/Yager,_Dexter_%26_Birdie.

US Department of Housing and Urban Development. *2015 AHAR: Part 1—PIT Estimates of Homelessness in the U.S.* Washington, DC: 2015.

———. *HUD 2015 Continuum of Care Homeless Assistance Programs Homeless Populations and Subpopulations, State Name: Florida*. Washington, DC: 2015.

US Department of Housing and Urban Development, US Department of Veterans Affairs, and US Interagency Council on Homelessness. *Mayors Challenge to End Veteran Homelessness in 2015 Fact Sheet*. Washington, DC: US Department of Housing and Urban Development, 2015.

US Department of Justice. "Justice Department Files Brief to Address the Criminalization of Homelessness." News release, August 6, 2015. https://www.justice.gov/opa/pr/justice-department-files-brief-address-criminalization-homelessness.

Van Sickler, Michael. "St. Petersburg City Council Passes Street Solicitation Ban." *Tampa Bay Times*, June 3, 2010. http://www.tampabay.com/news/politics/local/st-petersburg-city-council-passes-street-solicitation-ban/1099765.

Wealth Generators. "Dexter Yager." WallStreetGenerators.com, April 18, 2016. http://wallstreetgenerators.com/dexter-yager-the-amway-legend/.

Wikipedia. s.v. "Amway." Last modified May 20, 2016. https://en.wikipedia.org/wiki/Amway.

———. s.v. "Mary Baker Eddy." Last modified February 20, 2016. https://en.wikipedia.org/wiki/Mary_Baker_Eddy.

"Wildlife Rehabilitation Permit," Rule: 68A-9.006, Fish and Wildlife Conservation Commission. Florida Administrative Code & Florida Administrative Register. Effective date July 1, 2013. https://www.flrules.org/gateway/RuleNo.asp?ID=68A-9.006.

Williams, Deborah. "World-Renowned Sanctuary Most Definitely for the Birds." *Sun-Sentinel*, March 25, 1990. http://articles.sun-sentinel.com/1990–03–25/features/9001310440_1_birds-heath-spread.

Willis, Derek. "To Understand Scott Walker's Strengths, Look at His Donors." *New York Times*, February 11, 2015. http://www.nytimes.com/2015/02/12/upshot/to-understand-scott-walkers-strength-look-at-his-donors.html?_r=0.

Windsor. "Windsor Fact Sheet." Accessed June 9, 2016. http://www.windsorflorida.com/app/uploads/2016/03/Windsor-Fact-Sheet.pdf.

Young, M. Scott, and Kathleen A. Moore. *2014 Point-in-Time Housing Survey*. St. Petersburg, FL: Pinellas County Homeless Leadership Board, 2014.

OTHER SOURCES

Anthony Catron, Raymond Young, Charles R. Hargis, et al. v. City of St. Petersburg. US Court of Appeals, Eleventh Circuit, 2011, no. 10–12032 658 F.3d 1260 1264.

Beitl, Adrianne. Email to the author, October 28, 2015.

———. Interview by the author, Suncoast Seabird Sanctuary, Indian Shores, FL, July 12, 2015.

Bolden, George. Interview by the author, (swah-rey), St. Petersburg, FL, January 30, 2016.

———. Welcoming statements, (swah-rey), St. Petersburg, FL, January 30, 2016.

Dale. Interview by the author, Bayou Club, Largo, FL, July 16, 2015.

DeChant, Dell. Letter to Emma Curtis Hopkins College provisional board, March 1, 1992.

———. Letter to Emma Curtis Hopkins College board, May 15, 1992.

———. Letter to Patricia Gerard, July 13, 1992.

———. Letter to Unity-Progressive Council higher education subcommittee, June 5, 1994.

———. Letter to Unity-Progressive Council higher education subcommittee, June 6, 1994.

Deed of sale from Zeta G. Bobbitt to Suncoast Seabird Sanctuary Inc., February 7, 1997 (filed February 11, 1997). Pinellas County, FL, deed book 9606, page 1237.

Divorce certificate, Sixth Judicial Circuit of Florida, Pinellas County, circuit civil no. 74–9383–07. Clearwater, FL, December 19, 1974.

Doty, Kevin S. Letter about Ralph Heath, October 29, 2015.

Emma Curtis Hopkins College. Board meeting minutes, April 16, 1993.

———. Board meeting minutes, July 15, 1993.

———. Checklist for temporary licensure, July 1992.

Eslick, Micki. Email to Robin Vergara, June 14, 2012.

———. Interview by the author, Eslick's home, November 4, 2015.

Florida Fish and Wildlife Conservation Commission. Inspection report, April 30, 2014, Pinellas County, FL, ref. no. FWSW14OFF5258.

———. Inspection report, June 17, 2014. Pinellas County, FL, ref. no. FWSW140FF7972.

———. Inspection report, May 5, 2016, Pinellas County, FL, ref. nos. FWSW16OFF006622 and FWTB16CAD012948.

Gerard, Eric. Interviews by the author, phone, March 20, April 11, and May 12, 2016.

Gerard, Patricia. Interview by the author, home of Patricia Gerard, Largo, FL, July 11, 2015.

———. Interviews by the author, phone, March 15, March 28, and May 8, 2016.

———. Letter to Emma Curtis Hopkins College board, July 2, 1993.

———. Prayer journal, 1992.

Godwin, Jennifer. Interview by the author, home of Robin Vergara, Indian Rocks Beach, FL, July 24, 2015.

Gregory, Lar. Interview by the author, phone, November 29, 2015.

Guastella, Jimbo. Interview by the author, phone, November 27, 2015.

———. Interview by the author, Suncoast Seabird Sanctuary, Indian Shores, FL, August 3, 2015.

Hammock, Rev. Leddy. Email to the author, March 14, 2016.

———. Interviews by the author, Unity Church of Clearwater, FL, July 5 and August 12, 2015.

———. Letter of invitation to Emma Curtis Hopkins College opening, July 6, 1995.

———. "Rejuvenate." Sermon, Unity Church of Clearwater, FL, July 5, 2015.

———. Welcoming invocation of Emma Curtis Hopkins College, Summer 1995.

Heath, Ralph. Interviews by the author, Greek Village Restaurant, Seminole, FL, November 2, 4, 2015, and Salty's, Gulfport, FL, July 30, 2015.

———. Interviews by the author, phone, August 1, August 12, and June 5, 2015.

Kraut, Kim, and Greg Vaughan. Interview by the author, Suncoast Seabird Sanctuary, Indian Shores, FL, July 29, 2015.

Kriseman, Rick. "State of the City Address." Address delivered to the City of St. Petersburg, FL, January 24, 2015.

Los Angeles County Sheriff's Department, Inmate Information Center, booking no. 4479724. Los Angeles, CA, 2015. https://app4.lasd.org/iic/details.cfm.

Marbut, Robert. "Follow-Up Review of Homelessness in the City of St. Petersburg: Presentation of Findings and Action Plan Recommendations to the City of St. Petersburg." Report presented to the City of St. Petersburg, FL, June 5, 2014.

———. "Homeless Needs Assessment and Action Plan for Placer County." Report presented to Placer County, CA, April 7, 2015.

Marler, George. Email to the author, November 6, 2015.

Marriage certificate, Commonwealth of Virginia Department of Health, Division of Vital Records, file no. 88–008792. Williamsburg, VA, March 7, 1988.

Megan. Interview by the author, (swah-rey), St. Petersburg, FL, January 30, 2016.

Notice of bankruptcy case filing, US Bankruptcy Court, Middle District of Florida, case number 8:13-bk-1315Q-CED. Tampa, FL, October 1, 2013.

Osmundson, Linda. Interview by the author, Osmundson's home, St. Petersburg, FL, July 13, 2015.

Pavese, Irene. Interview by the author, (swah-rey), St. Petersburg, FL, January 30, 2016.

Pinellas County deputy. Interveiw by the author, Safe Harbor, Clearwater, FL, July 30, 2015.

Raposa, Michael. Interview by the author, St. Vincent de Paul, St. Petersburg, FL, January 25, 2016.

Realtor 1. Interview by the author, Silverthorn Road, Seminole, FL, July 16, 2015.

Realtor 2. Interview by the author, Eagle Pointe Drive, Clearwater, FL, August 8, 2016.

Realtor 3. Interview by the author, Bullard Drive, Clearwater, FL, August 8, 2015.

Rolle, G.W. Interviews by the author, phone, January 25 and January 28, 2016.

———. Interviews by the author, Rolle's home, St. Petersburg, FL, November 3, 2015; Trinity Lutheran Church, St. Petersburg, FL, July 11, July 17, and August 8, 2015, and January 30, 2016; and Williams Park, St. Petersburg, FL, January 24, 2016.

———. Text messages to the author, February 4 and February 19, 2016.

———. "Things That Break." Unpublished manuscript, 2015.

Ronald J. Cooper vs. Suncoast Seabird Sanctuary, et al. Civil Court of Pinellas County, FL, Sixth Circuit, 2013. Ref. no. 13–710 CI-8. 18277, 632.

Smith, Clifford. Interview by the author, city hall, St. Petersburg, FL, January 25, 2016.

Snapp, Tom. Interview by the author, Trinity Lutheran Church, St. Petersburg, FL, July 19, 2015.

Spicer, H. Jane. Letter to Emma Curtis Hopkins College board, May 21, 1996.

Unity. "Daily Word." July 15, 2015.

United Church Directories, *Unity-Clearwater Church, Clearwater, Florida.* 1994.

Unity-Clearwater Church. *Golden Anniversary: 50 Years of Unity-Clearwater 1941–1991*, anniversary booklet, 1991.

Unity-Progressive Council, Inc. *A Progressive Reaffirmation of Unity Faith.* Clearwater, FL: Unity-Progressive Council, Inc., 1990.

Unknown author. Letter of response to Alan Rowbotham, June 1994.

Vaughan, Greg. Interview by the author, phone, October 28, 2015.

———. Interviews by the author, Suncoast Seabird Sanctuary, Indian Shores, FL, July 12 and August 3, 2015.

Vaughan, Greg, and Kellie Vaughan. Interviews by the author, Salty's, Gulfport, FL, July 30, 2015, and the Vaughans' home, Largo, FL, August 10, 2015.

Vergara, Robin. Email to the author, November 10, 2015.

———. Email to von Gontard brothers, August 2, 2012.

———. Interview by the author, Vergara's home, July 24, 2015.

Vickery, Shelley. Interview by the author, Suncoast Seabird Sanctuary, Indian Shores, FL, July 12, 2015.

Walls, Chris. Interview by the author, Suncoast Seabird Sanctuary, Indian Shores, FL, July 14, 2015.

Whitaker, Bob, and Campbell Whitaker. Letter to members of the Emma Curtis Hopkins College steering committee, June 29, 1994.

Whitney, Ruth. Interview by the author, phone, March 31, 2016.

Wiggins, Carolyn. Email to the author, November 24, 2015.

ENDNOTES

Mother-Father God

18 *Leddy told the story of the disciples asking Jesus to hush the children*: Hammock, interview, March 14, 2016.

22 *One night in 1978, my mom went to a biker bar in South Tampa to hear the Mad Beach Band*: P. Gerard, interview, July 11, 2015.

23 *One day, she was supposed to meet the man who was installing new carpet*: Ibid.

24 *Leddy Hammock had taken over as co-minister just the year before*: Unity-Clearwater Church, *Golden Anniversary.*

25 *invited my mom to meet them, too*: P. Gerard, interview, March 15, 2016.

25 *Increased industrialization was reducing the demand for men's physical labor*: Satter, *Each Mind a Kingdom*, 8.

25 *inspired the rise of moral reform groups largely led by women*: Ibid., 22.

25 *The number of women in higher education also increased dramatically, as did the number of women in the workforce*: Ibid., 21.

25 *research in the scientific and medical fields was raising questions about the unseen world*: Ibid., 49–51.

26 *Germ theory, popularized in 1876, introduced the general populace to the theory that sickness was caused by outside forces, not internal weaknesses*: Ibid., 51–52.

26 *The debate gave new, scientific validation to more general fears of contagion*: Ibid., 51.

26 *Many women saw hypnotism as an opportunity to bring about a new era in which mind ruled matter*: Ibid., 52.

26 *their claims to power had previously been based upon self-sacrifice and spiritual superiority on the grounds of moral purity, reinforcing the association of women with the "feminine heart"*: Ibid., 30–31.

26 *Mary Baker Eddy traveled from Rumney, New Hampshire, to Belfast, Maine, to see a famous healer named Phineas Parkhurst Quimby*: Ibid., 62.

27 *Quimby had reportedly cured thousands of people around New England*: Ibid., 60.

27 *Eddy was the youngest of six children*: Rogers, "Mary Baker Eddy and the American Dream."

27 *a quick-tempered and punishing father*: *Wikipedia*, s.v. "Mary Baker Eddy."

27 *suffered from a nervous sickness that resulted in fainting episodes*: Rogers, "Mary Baker Eddy and the American Dream."

27 *She treated chronic indigestion with a strict diet of water, vegetables, and bread*: *Wikipedia*, s.v. "Mary Baker Eddy."

27 *Eddy had turned to allopathic medicine and alternative therapies such as homeopathy and hydrotherapy to treat her mysterious illnesses*: Rogers, "Mary Baker Eddy and the American Dream."

27 *Eddy's first husband died suddenly in 1844, when she was six months pregnant*: *Wikipedia*, s.v. "Mary Baker Eddy."

27 *she became interested in mesmerism, animal magnetism, Spiritualism, clairvoyance, and the stories of Jesus as a healer*: *New World Encyclopedia*, s.v. "Mary Baker Eddy."

27 *Eddy's second husband promised to become the boy's legal guardian*: Rogers, "Mary Baker Eddy and the American Dream."

27 *he never followed through, and Eddy lost touch with her son for the next twenty years*: *Wikipedia*, s.v. "Mary Baker Eddy."

27 *Her new husband was unfaithful . . . he would eventually abandon her*: "The Life of Mary Baker Eddy," *Mary Baker Eddy Library.*

28 *For the next three years, Quimby tutored Eddy in his brand of mesmerism, and they kept in close touch until his death in 1866*: *New World Encyclopedia*, s.v. "Mary Baker Eddy."

28 *To treat these cases, he would furnish his patients with reinterpretations of Bible stories he'd penned himself*: Satter, *Each Mind a Kingdom*, 60.

28 *What appears to exist as the material world is an illusion caused by another creative force: "mortal mind," or human thought*: Ibid., 62.

28 *It believes in evil and, by believing in it, manifests it in the illusionary world*: Ibid., 63.

28 *as children of Divine Mind, humans have godlike abilities and can overcome the appearances of evil*: Ibid.

28 *One does this by silently "arguing" away the appearance of evil*: Ibid., 64.

29 *in Christian Science the evil thoughts of others make us sick*: Ibid., 65.

29 *Eddy was giving lectures to packed theaters around the northeast*: Ibid., 1.

29 *That September, Eddy appointed her editor of the fledgling* Christian Science Journal: Ibid., 81.

29 *my mom went to work for the Largo Police Department in Largo, Florida, in 1984*: P. Gerard, interview, July 11, 2015.

29 *they hadn't had a victim advocate in months*: P. Gerard, interview, March 28, 2016.

29 *One afternoon, a client wandered away from his group home and was picked up by police*: P. Gerard, interview, July 11, 2015.

30 *Florida legislators toughened the state's domestic violence laws, enabling police to make arrests on misdemeanor assault or battery charges even if they hadn't witnessed an incident*: Gibeaut, "Domestic Violence Arrests May Slow Vicious Cycle."

30 *at the time, there was no state- or countywide protocol for responding to domestic violence calls*: P. Gerard, interview, July 11, 2015.

30 *there wasn't even a domestic violence detective on the Largo police force*: P. Gerard, interview, March 28, 2016.

30 *Officers would show up, walk the offender around the block, then take him home*: P. Gerard, interview, July 11, 2015.

30 *victims recanted their claims, which seemed like a waste of everyone's time to the officers*: P. Gerard, interview, March 28, 2016.

30 *Florida legislature passed a law enabling victims to take out injunctions for protection against their abusers*: Phillips, "Spouse Abuse Laws Helping Victims."

31 *law enforcement generally considered matters between spouses to be private*: P. Gerard, interview, July 11, 2015.

31 *When officers responded to a rape call, or a child sexual abuse call, or a domestic violence call, they were supposed to call my mom to the scene*: P. Gerard, interview, March 28, 2016.

31 *"I'm your victim advocate! Use me!"*: P. Gerard, interview, July 11, 2015.

32 *That year, the local chapter of the National Organization for Women (NOW) began lobbying the Police Standards Council*: Whitney, interview, March 31, 2016.

32 *in 1986, the county passed the policy*: P. Gerard, interview, March 28, 2016.

32 *Soon after, the Standards Council formed a Domestic Violence Task Force*: Whitney, interview, March 31, 2016.

33 *She had returned to a puritanical religious approach in Christian Science, eschewing all other schools of metaphysics*: Ibid., 15–16.

33 *She considered herself something of a prophet with a sole claim to truth*: Satter, *Each Mind a Kingdom*, 82.

33 *In the year since her editorship began, Hopkins had shown herself to be an important actor in the Christian Science movement*: Harley, *Emma Curtis Hopkins*, 15–18.

33 *In 1885, thirteen months after hiring Hopkins, Eddy abruptly dismissed her*: Satter, *Each Mind a Kingdom*, 81.

33 *With encouragement from Plunkett, in late 1885 Hopkins left her husband and son to move to Chicago*: Ibid., 82.

33 *By spring 1886, she'd founded the Emma Curtis Hopkins College of Christian Science*: Harley, *Emma Curtis Hopkins*, 38.

33 *she founded, alongside a group of prominent students, the Hopkins Metaphysical Association*: Satter, *Each Mind a Kingdom*, 82.

34 *by the end of the following year, between seventeen and twenty-one Hopkins Metaphysical Associations were operating across the country*: Ibid.; Harley, *Emma Curtis Hopkins*, 51–52.

34 *By the end of 1887, she had personally instructed six hundred students*: Satter, *Each Mind a Kingdom*, 82.

34 *Hopkins taught that God is all, God is good, and God is Mind*: Ibid., 86.

34 *Hopkins strongly opposed teachings about sin and repentance*: Ibid.

34 *Hopkins taught them to "enter the silence" and meditate on "affirmations" and "denials"*: Ibid., 87.

34 *"I AM power of Life to the universe"*: Newman, "The Radiant I Am by Emma Curtis Hopkins (Animated)."

35 *In Hopkins's ideology, God is Father, Son, and Mother-Spirit, or "Holy Comforter"*: Melton, "Emma Curtis Hopkins: A Feminist of the 1880s and Mother of New Thought," 93–94.

35 *1986, the year Pinellas County passed its Preferred Arrest policy*: Whitney, interview, March 31, 2016.

35 *they could rip up their injunction in front of a police officer and the officer couldn't do anything about it*: P. Gerard, interview, July 11, 2015.

35 *There were no domestic violence units in state attorneys' offices, and no protocol for prosecuting abusers if victims refused to testify*: Neal, "Agencies Rated Poorly on Response to Abuse Cases."

37 *Plunkett had absconded to New York with the mailing list for* Truth: Harley, *Emma Curtis Hopkins*, 44.

37 *Hopkins transformed the College of Christian Science into the Christian Science Theological Seminary*: Ibid., 83.

37 *deemphasizing the professional aspect of Christian Science and emphasizing ministry*: Ibid., 44.

37 *she met one-on-one with every advanced student in his or her final year, developing an individualized curriculum for each*: Ibid., 46.

37 *The seminary's first ordination ceremony was held in January 1889*: Satter, *Each Mind a Kingdom*, 84.

38 *"Divine Truth has come at last to give woman her proper status in the world"*: Harley, *Emma Curtis Hopkins*, 84.

38 *In the seminary's lifetime, Hopkins would ordain hundreds of female ministers*: Ibid., 64.

38 *Myrtle Fillmore attended a mental-healing seminar in Kansas City, Missouri, held by the New Thought teacher E. B. Weeks*: Freeman, *The Story of Unity*, 44–45.

38 *Myrtle's symptoms flared and her doctors recommended she leave Kansas City for preferable climes*: Ibid., 40.

38 *"I am a child of God and therefore I do not inherit sickness"*: Ibid., 45.

38 *attending every metaphysical lecture that came through Kansas City*: Satter, *Each Mind a Kingdom*, 105.

38 *The following year, she and her husband, Charles, began publishing their own New Thought journal,* Modern Thought: Ibid., 106.

38 *in 1891, Hopkins ordained them at the Christian Science Theological Seminary*: Ibid., 107.

39 *It was the heyday of New Thought, and churches were springing up in diffuse locations across the country*: Harley, *Emma Curtis Hopkins*, 64.

39 *Myrtle supplied the original impetus for founding what was then called the Unity School of Practical Christianity*: Ibid., 108.

39 *since childhood he had immersed himself in Shakespeare, Tennyson, Lowell, and Emerson*: Freeman, *The Story of Unity*, 27.

39 *works of Spiritualism, the occult, and Eastern religion*: Satter, *Each Mind a Kingdom*, 107.

39 *Charles had rarely attended church as a child*: Freeman, *The Story of Unity*, 23.

39 *Fillmores taught that God is all, God is good, and God is Mind*: deChant, "Myrtle Fillmore and Her Daughters," 106.

39 *Affirmation and denial are cornerstones of the faith, as is the application of such statements in healing*: Satter, *Each Mind a Kingdom*, 109.

39 *some early pamphlets and articles published by Unity feature titles such as "Curing Colds," "An Airplane Blessing," "An Automobile Blessing," and "A Salesman's Prayer"*: Braden, *Spirits in Rebellion*, 245–246.

40 *we are all Divine. We can all do what Jesus did*: Ibid., 261.

40 *At the insistence of Mary Baker Eddy, the Fillmores abandoned the name Christian Science*: Ibid., 254.

40 *adopted that of the Unity School of Christianity*: Ibid., 252.

40 *it was the Fillmores' intention to establish not a church*: Ibid., 239.

40 *Charles finally sought to make clear to himself and others what he really thought and believed*: Ibid., 252.

40 *he did so in a set of twelve informal lessons based on his own personal experiences, which he taught in an informal, discussion-based format over two weeks*: Ibid.

40 *Unity finally came to be formalized as a church*: Ibid., 253.

40 *Now people were being taught the principles of Unity from a definite point of view, and going on to establish churches of their own bearing the Unity name*: Ibid.

40 *a Unity field department was established and Unity ministers naturally organized themselves into a ministers association*: Ibid., 254.

40 *any group that wished to be recognized had to adhere to the Unity teachings and textbooks*—and: *get rid of any texts and teachings that didn't conform to the Christ Standard*: Ibid., 255.

41 *It was expected that the standards would be raised with the passage of time*: Ibid., 257.

41 *"in prayerful openness to the Spirit of Truth and with the guidance of God"*: deChant, *Unity and History*, 48.

41 *Concern had arisen among them that the Unity religion was becoming diluted by Unity churches whose practices were not in line with its foundational teachings*: E. Gerard, interview, April 11, 2016.

42 *Lowell was a religious moderate and had a vision of growing Unity's reach in America*: deChant, *Unity and History*, 40.

42 *allowing Unity churches to become autonomous*: Ibid., 42.

42 *promoting practices like crystal healing and channeling, or conversing with spirits*: Ibid., 46.

42 *practices Charles Fillmore had specifically spoken against*: E. Gerard, interview, April 11, 2016.

42 *Then Charles Rickert passed and his daughter Connie Fillmore Bazzy assumed his place in 1987*: deChant, *Unity and History*, 41.

42 *Unity School had been a center for prayer, publication, and ministerial training*: Ibid., 42.

43 *the AUC had come about in response to a perceived crisis in doctrine and a failure of leadership*: Ibid., 33.

43 *Unity School, which had come to have little to do with the operation of Unity churches*: Ibid., 43.

43 *Now the AUC served as the organizing body of as many as five hundred Unity ministers*: Ibid.

43 *Lowell's revisionist leadership had caused a de-emphasis in the movement on such central concepts*: deChant, *Unity and History*, 40.

43 *Whereas there was intellectual commerce between the AUC and Unity School, they were functionally and philosophically independent*: deChant, "Myrtle Fillmore and Her Daughters," 104.

43 *This gulf between the two leading Unity institutions resulted in a contradiction, and resultant vagueness, of self-definition*: deChant, *Unity and History*, 46.

43 *Its prayer center, Silent Unity, was receiving over a million requests for prayer annually*: Freeman, *The Story of Unity*, 220.

43 *"What is Unity?"*: deChant, *Unity and History*, 46.

44 *Graduates from the Unity Ministerial School and the Unity School of Religious Studies weren't well versed in Unity's teachings or its history*: Ibid., 47.

44 *Unity didn't even have an accredited theological seminary at the time*: Ibid.

44 *students in Unity schools were being taught by teachers who lacked terminal academic degrees*: Ibid.

44 *the U-P.C. had attempted to reach out to Unity headquarters in Unity Village, Missouri, with their concerns, they were rebuffed*: E. Gerard, interview, March 20, 2016.

44 *The U-P.C. proceeded on the basis of three affirmations*: deChant, *Unity and History*, 48.

45 *Those wishing to be members of the U-P.C. are asked to sign their names at the bottom*: Unity-Progressive Council, Inc., *A Progressive Reaffirmation of Unity Faith.*

46 *"I was able to thank the person I had been focused on last night for the things she had given me: self-awareness, a clearer sense of my role in my job"*: P. Gerard, prayer journal.

47 *"amazed to realize how angry I have been and how dramatically different that feels right now"*: Ibid.

47 *"We're all really practicing the same religion"*: Ibid.

47 *"I emotionally felt the Presence, and just let it happen. Joy!"*: Ibid.

48 *"I always end up stronger and clearer-minded when the dust settles, but it's most disturbing while it's going on"*: Ibid.

49 *the Florida Legislature ordered all twenty state attorneys' offices to establish domestic violence units*: Neal, "Agencies Rated Poorly on Response to Abuse Cases."

49 *the Florida Supreme Court ordered district courts to institute family law divisions*: Hallifax, "Justices Rap Domestic Violence Law."

49 *Dell invited my mom to serve on the provisional board of directors*: deChant, letter to Emma Curtis Hopkins College provisional board, March 1.

49 *The plan was to develop a new division of the Unity-Progressive Theological Seminary into an entirely independent educational institution*: Ibid.

50 *The first meeting convened on June 30 at Leddy's home*: deChant, letter to Emma Curtis Hopkins College board, May 15, 1992.

50 *Dell had already begun the process of laying out degree requirements and a course catalog*: Emma Curtis Hopkins College, checklist for temporary licensure.

50 *It was decided that my mom would be president of the board and chair the steering committee*: deChant, letter to Patricia Gerard, July 13, 1992.

50 *Emma Curtis Hopkins College board met late in the evening at the home of one of its members to discuss alternatives for proceeding with the opening of the school*: Emma Curtis Hopkins College, board meeting minutes, April 16, 1993.

51 *In July, they reconvened at my home*: P. Gerard, letter to Emma Curtis Hopkins College board, July 2, 1993.

51 *Of the $30,000 they'd determined to raise in April, just $12,000 had been pledged and $3,000 collected*: Emma Curtis Hopkins College, board meeting minutes, July 15, 1993.

53 *"a hostile and negative series of announcements"*: Unknown author, letter of response to Alan Rowbotham, June 1994.

54 *"This is not a personal matter to us"*: Ibid.

54 *decided to move forward with incorporating the school and preparing its application materials for the Florida Department of Education*: deChant, letter to the U-P.C. higher education subcommittee, June 5, 1994.

54 *any gifts, financial or otherwise, that the U-P.C. higher education subcommittee desired to share, especially books*: deChant, letter to the U-P.C. higher education subcommittee, June 6, 1994.

54 *By that time, the board had collected just $8,000*: deChant, letter to the U-P.C. higher education subcommittee, June 5, 1994.

54 *Bob and Campbell Whitaker, president and dean of the school, came to the board with grave news*: Whitaker and Whitaker, letter to Emma Curtis Hopkins College steering committee, June 29, 1994.

54 *In the church directory from that year, I appear in one photo*: *United Church Directories*, Unity-Clearwater Church, 1994.

56 *The opening ceremony of the Emma Curtis Hopkins College was held at a sunny six in the evening in August 1995*: Hammock, letter of invitation, July 6, 1995.

56 *By this time, the congregation of Unity-Clearwater had grown to almost 1,500 members*: Unity-Clearwater Church, *Golden Anniversary*.

57 *"Dear friends, we bless and praise the grand opening of the Emma Curtis Hopkins College here this beautiful Indian summer"*: Hammock, welcoming invocation.

57 *work on the library was diligent and accelerating*: Spicer, letter to Emma Curtis Hopkins College board, May 21, 1996.

62 *"I am whole, strong, and full of vitality"*: Unity, "Daily Word," July 5, 2015.

63 *"How do we get to be a family?"*: Hammock, "Rejuvenate," July 5, 2015.

63 *"There are no accidents"*: Ibid.

65 *"We survived in spite of the fire"*: Hammock, interview, July 5, 2015.

66 *"God said let there be light, and there was light"*: Hammock, interview, August 12, 2015.

66 *Cady was a homeopathic physician in New York when Myrtle Fillmore discovered her pamphlet*: Braden, *Spirits in Rebellion*, 244.

66 *Myrtle and Charles later collected Cady's writings as* Lessons in Truth: Ibid., 244–245.

66 *Maya Angelou once called it a revelation on* The Oprah Winfrey Show: Chapman, "Maya Angelou: 'God Loves Me'—'That's Why I Am Who I Am.'"

66 *"Seek light from the Spirit of Truth within you. Go alone. Think alone"*: Cady, *Lessons in Truth*, 32.

67 *"The sun does not radiate light and warmth today and darkness and chill tomorrow"*: Ibid., 39.

67 *"We do not have to beseech God any more than we have to beseech the sun to shine"*: Ibid., 38.

67 *"If you repeatedly deny a false or unhappy condition, it loses its power to make you unhappy"*: Ibid., 50.

67 *"It is perfectly natural for the human mind to seek to escape from its troubles by running away from present environments or by planning some change on the material plane"*: Ibid., 12.

67 *"You need take no part in the outer demonstration of relief from conditions"*: Ibid.

68 *"Inharmony cannot remain in any home where even one member daily practices [an] hour of the presence of God"*: Ibid., 10.

69 *"Unfortunately, my faith kept me in the relationship too long"*: Osmundson, interview, July 13, 2015.

69 *"I don't know if it's possible"*: Osmundson, interview, July 13, 2015.

Going Diamond

74 *Some call it a pyramid scheme*: O'Donnell, "Multilevel Marketing or 'Pyramid?' Sales People Find It Hard to Earn Much."

74 *claimed transglobal sales of $9.5 billion*: Martinez, "Amway's 2015 Revenues Fall to Lowest Level in 5 Years."

74 *It is the biggest direct-selling company in the world*: Coffey and de Jong, "Amway Billionaires Debut After Bad Year for Direct Seller."

74 *"selling" Amway products*: Roth, "Dreams Incorporated."

74 *who then distribute Amway products to other distributors they sign up*: Amway, *Business Reference Guide*.

74 *a second year of decline for Alticor, whose sales in 2014 totaled $10.8*

billion: Coffey and de Jong, "Amway Billionaires Debut After Bad Year for Direct Seller."

75 *multimillion-dollar lawsuits and other legal actions on almost every continent*: *Wikipedia*, s.v. "Amway."

75 *After all, less than 1 percent of Amway distributors go Diamond*: Amway, *Business Reference Guide*; Pyramid Scheme Alert, "Study of Ten Major MLMs and Amway/Quixtar."

75 *all yards are maintained by the Bayou Club's landscapers*: Realtor 1, interview, July 16, 2015.

78 *when they commissioned him to write the first "official" history of the Amway Corporation*: Cross, *Choices with Clout*, xi.

78 *change negative self-talk into positive self-talk*: Cross, *Amway*, 46.

78 *if they can't train them to be positive*: Ibid., 100–102.

79 *a central reason why Amway is so successful*: Ibid., 99–103.

79 *"a skill that has to be learned, practiced, and put into action"*: Ibid., 99.

79 *"practice the art . . . on a daily basis"*: Ibid.

80 *"Most people think the kitchen needs updating"*: Realtor 1, interview, July 16, 2015.

82 *Rich was fourteen*: DeVos, *Simply Rich*, 31.

82 *He was walking two miles through the snow to his high school*: Ibid., 27.

82 *"already being an enterprising type, I had an idea"*: Ibid., 28.

83 *"I thought a ride in this car would surely beat the bus, a streetcar, or walking"*: Ibid.

83 *running a pilot school*: Ibid., 50.

83 *comedy-of-errors trip on a sailboat*: Ibid., 58–69.

83 *for which Jay's second cousin and his parents are already distributors*: Ibid., 74.

83 *"Until then, there had been no official government position on what type of claims could be made about dietary supplements"*: Ibid., 85.

83 *let nothing stand in your way, then success is guaranteed*: Cross, *Amway*, 45.

83 *Rich DeVos calls "Compassionate Capitalism"*: DeVos, *Compassionate Capitalism*.

83 *Amway's first original product*: DeVos, *Simply Rich*, 97.

83 *investigation of Amway, by the Federal Trade Commission in 1975*: Ibid., 126.

84 *"misunderstanding of business principles and an attack on free enterprise"*: Ibid., 131.

84 *challenges free enterprise, and thus freedom itself*: Cross, *Amway*, 94–99.

84 *"I wish they would take responsibility for their own actions instead of trying to blame the business"*: DeVos, *Simply Rich*, 130.

84 *they lack faith in their ability to succeed, and thus the necessary determination*: Cross, *Amway*, 86–87.

84 *millions worldwide*: DeVos, *Simply Rich*, 153–154.

84 *YouTube video uploaded in 2011*: "Amway Function," YouTube video.

85 *"All it took was the willingness to work hard to achieve a dream"*: DeVos, *Simply Rich*, 87.

86 *"This is Renata, my assistant"*: Realtor 2, interview, August 8, 2015.

90 *"This would make a good nursery"*: Ibid.

92 *"Listen to Alcimon and Marie-Chantale Colas"*: Cross, *Amway*, 6.

92 *"Listen to Sevgi Corapci"*: Ibid., 3.

92 *"I was a salaried man working in a company for eight years"*: Ibid.

93 *"they noticed I became more optimistic and more healthy"*: Ibid.

93 *Listen to Rosemarie and Otto Steiner-Lang*: Ibid.

93 *beating boredom and passivity, spending more time with family, and earning the admiration of others*: Ibid., 5.

93 *A housewife can do it*: Ibid.

93 *The physically enfeebled can do it*: Ibid., 36.

94 *pioneering move onto the World Wide Web*: Ibid., 69–81.

94 *"Amway Distributor Profile"*: Ibid., 172–192.

94 *"Bootstraps Philosophy"*: Ibid., 13–33.

94 *Amway's foreign expansion strategy*: Ibid., 135–153.

94 *"sell products and keep on generating volume"*: Ibid., 181.

94 *Not in retail stores*: Amway, *Business Reference Guide*, D-13–D-14.

94 *Amway distributors can only sell their products directly to the public, or to other Amway distributors*: Ibid., D-4.

94 *in blind tests, Amway products often score poorly*: O'Donnell, "Multilevel Marketing or 'Pyramid?'"

94 *"putting free enterprise in the hands of the common man and woman"*: Cross, *Amway*, 8.

95 *making any sacrifices necessary*: Ibid., 113–115.

95 *Amway was bigger than making money; it was a way to overhaul your lifestyle and live your dreams*: Ibid., 121.

95 *Yager, who didn't let a stroke stop him*: "Yager, Dexter & Birdie," Unofficial Amway Wiki.

95 *even as he was learning to walk and speak again*: Cross, *Amway* 121–122.

96 *"just performance"*: Ibid., 108–110.

96 *"Tastewise, it may not be your taste"*: Realtor 3, interview, August 8, 2015.

104 *Negativity causes failure*: Cross, *Amway*, 91.

104 *Cross spent the ten years after writing* Commitment to Excellence *researching the two men*: Cross, *Choices with Clout*, xi.

104 *"Money—and what it can buy—is the universally recognizable indicator of success"*: Cross, *Amway*, 79.

105 *"Are you ready now?"*: Cross, *Amway*, 101–102.

105 *yourself, family, community, free enterprise, America, and faith itself*: Ibid., 86.

105 *it doesn't endorse a single religion*: DeVos, *Simply Rich*, 217–218.

105 *people of all races, political affiliations, and creeds succeeding in Amway*: Ibid.

106 *recruiting more distributors*: Cross, *Amway*, 87.

106 *persisting in The Business through every hardship*: Ibid., 111–112.

106 *Free enterprise is a blessing from God*: DeVos, *Simply Rich*, 217.

106 *In 2016, Forbes estimated Rich DeVos's net worth to be $4.8 billion*: "The World's Billionaires," *Forbes*.

106 *you'd never do anything more than take the bus*: DeVos, *Simply Rich*, 186–187.

107 *$1 million renovation*: "Welcome to The Bayou Club," Bayou Club.

107 *new roof*: Dale, interview, July 16, 2015.

107 *"even among the ultradwarf cultivars, there is no other grass capable of producing the incredible ball roll of a well-maintained Champion green"*: "Champion: The Ultimate Bermudagrass Putting Surface," Champion Turf Farms.

107 *He tells us the club no longer has an initiation fee*: Dale, interview, July 16, 2015.

110 *lifetime charity donations of $1.2 billion*: "America's 50 Top Givers," *Forbes*.

110 *Most of that money has stayed in West Michigan*: Martinez, "$1.2 Billion in Donations Puts DeVos Family in Forbes Top Philanthropy List."

110 *have bequeathed a considerable portion of Grand Rapids*: DeVos, *Simply Rich*, 221–231.

111 *often credited for catalyzing the revitalization of downtown*: Helderop, "Detroit Can Learn a Thing or Two from Grand Rapids."

111 *Much of it went to public schools and Grand Rapids–based hospitals, arts programs, and faith-based organizations providing services to the homeless*: Martinez, "How and Why the DeVos Family Gives Away Billions."

111 *That same year over $4 million of DeVos money went to Hope College*: "Where DeVos Family Donates Millions," Mlive.com.

111 *Reformed Church in America—in which Rich DeVos was raised*: DeVos, *Simply Rich*, 213–214.

111 *while $2.2 million went to Calvin College, associated with the Christian Reformed Church in North America*: "Where DeVos Family Donates Millions," Mlive.com.

111 *13 percent went to churches and faith-based organizations*: Martinez, "How and Why the DeVos Family Gives Away Billions."

111 *$1.05 million to the Chicago-based Willow Creek Community*: Where DeVos Family Donates Millions," Mlive.com.

111 *"The Christian church and Christian education are high on our list of giving"*: DeVos, *Simply Rich*, 278.

111 *"My giving it puts the money in better hands than the government's"*: Ibid.

112 *Donors who meet at The Gathering dispense upwards of $1 billion a year in grants*: Michaelson, "The $1-Billion-a-Year Right-Wing Conspiracy You Haven't Heard Of."

112 *earning comparisons to the Kochs*: Kroll, "Meet the New Kochs."

112 *candidates like Newt Gingrich, Rick Santorum*: Bennett, "The Ultra-Rich, Ultra-Conservative DeVos Family."

112 *Jeb Bush*: Reens, "DeVos Family Goes All-In on Marco Rubio GOP Presidential Race."

112 *Scott Walker*: Willis, "To Understand Scott Walker's Strengths, Look at His Donors."

112 *Marco Rubio*: Reens, "DeVos Family Goes All-In on Marco Rubio GOP Presidential Race."

112 *ultraconservative organizations like Focus on the Family*: Bennett, "The Ultra-Rich, Ultra-Conservative DeVos Family."

112 *and the Family Research Council*: Kroll, "Meet the New Kochs."

112 *Alliance Defending Freedom*: Martinez, "How and Why the DeVos Family Gives Away Billions."

112 *the right's preeminent legal defense fund*: Michaelson, "The $1-Billion-a-Year Right-Wing Conspiracy You Haven't Heard Of."

112 *the Heritage Foundation*: "Where DeVos Family Donates Millions," Mlive.com.

112 *DeVos family donations accounted for over half of those made to the Michigan Republican Party*: "DeVos Family Responsible for HALF of Campaign Contributions to Michigan House Republicans in Last Quarter of 2015," *Eclectablog*.

113 *made an unsuccessful run for Michigan governor in 2006*: Bennett, "The Ultra-Rich, Ultra-Conservative DeVos Family."

113 *and now chairs the board of directors of the American Federation for Children*: "About," Dick and Betsy DeVos Family Foundation.

113 *appointed DeVos as finance chairman of the Republican National Committee*: DeVos, *Simply Rich*, 237–238.

114 *"which is a pretty solid Christian principle"*: Crawford, "What About NBA's Homophobic Owner?"

114 *"We also thought the ads might further help Amway distributors recognize the importance of free enterprise to their success"*: DeVos, *Simply Rich*, 238.

114 *Dexter Yager used Amway's extensive voice mail system*: Burstein and Lauerman, "She Did It Amway."

114 *"strong conservative"*: Bradsher, "And One Who Tries to Work Things Out."

114 *Republican "infomercials" airing on televangelist Pat Robertson's Family Channel*: Burstein and Lauerman, "She Did It Amway."

114 *"Master Dream Builder"*: Cross, *Amway*, 92.

114 *Yager's business now employs four million Amway team members in over forty countries*: Wealth Generators, "Dexter Yager."

114 *"Yager System," one of the first and most profitable motivational "tools" businesses run by Amway distributors*: Kneale and Lambert, "Climb to the Top."

115 *the Yager Group is still today an Amway-approved training provider*: "Amway Approved Provider—Yager Group," Amway.

115 *"Admirers speak of him with reverence, as if his next plateau of Amway achievement were sainthood itself"*: Morrill and Stancill, "Amway the Yager Way."

115 *"We're a longer course"*: Dale, interview, July 16, 2015.

118 *My parents more or less broke even in Amway*: E. Gerard, interview, May 12, 2016.

118 *She believed it was a cult*: P. Gerard, interview, May 8, 2016.

118 *"She wasn't going to leave me"*: E. Gerard, interview, May 12, 2016.

120 *part of my parents' strategy for "showing The Plan" was that they didn't even tell people it was Amway*: Ibid.

121 *There's an equestrian center, tennis courts, a concierge, and a gun club*: "Windsor Fact Sheet," Windsor.

121 *the DeVoses, who own three houses here and spend eight weeks a year or more on the waterfront*: Abkowitz, "A Resort With Less Glitz and No Kitsch."

121 *"Windsor is so different from the rest of Florida"*: Ibid.

122 *In 2010, Amway reached a settlement reportedly valued at $100 million in a California class action lawsuit*: Grand Rapids Press Staff, "Lawyers Say Their $20 Million Payment Is Fair."

122 *more than doubling the number of professional trainers, such as the Yagers, across the country*: Knape, "Amway Agrees to Pay $56 Million, Settle Case Alleging It Operates as a 'Pyramid Scheme.'"

122 *the CEO of Amway India was arrested for fraud*: Stroud, "The Indian Express."

122 *the biggest Amway distributorship in the world*: Berman, "Inside the Amway Sales Machine."

122 *Amway business owners make closer to $200 a month*: Reynolds, "Amway's Claims Are Malarkey, Distributors Say."

The Mayor of Williams Park

174 *"an imperfect church of imperfect people inviting other imperfect people to find the perfect love"*: "What Is the Missio Dei?" Missio Dei.

174 *"I identify as the hands and feet of Christ"*: Rolle, interview, November 3, 2015.

175 *In 2015, the state of Florida was home to an estimated 35,900 homeless people*: US Department of Housing and Urban Development, *HUD 2015 Continuum of Care*, 1.

176 *It had the second-highest number of unsheltered homeless*: US Department of Housing and Urban Development, *2015 AHAR: Part 1—PIT Estimates of Homelessness in the U.S.*

176 *It had the highest number of homeless veterans*: Ibid.

176 *6,853 homeless people; 40 percent of them are children*: Santa Lucia, *Pinellas County Point in Time Homeless Report: 2015*, 4.

176 *Homeless Leadership Board*: "Mission Statement," Pinellas County Homeless Leadership Board.

178 *"I wrote about it in my book"*: Rolle, interview, November 3, 2015.

178 *"Most of the people on the bottom aren't intellectuals"*: Rolle, interview, January 24, 2016.

178 *St. Petersburg passed an ordinance banning the roadside sale of newspapers*: Van Sickler, "St. Petersburg City Council Passes Street Solicitation Ban."

179 *Instead, G.W. spent a night in Los Angeles County jail*: Los Angeles County Sheriff's Dept., Inmate Information Center, booking no. 4479724.

179 *Convention Against Torture*: National Law Center on Homelessness and Poverty, *Cruel, Inhuman, and Degrading*, 5.

179 *lack of shelter space*: Ibid., 11.

179 *available services for families in crisis*: Ibid., 12–14.

179 *lack of affordable housing*: Ibid., 7.

179 *disproportionate representation of nonwhite and disabled people within the U.S. housing-unstable population*: Ibid., 15–16.

179 *local ordinances, such as those against panhandling and sleeping outside, which target unhoused people*: Ibid., 6.

180 *sharply on the rise since 2009:* National Law Center on Homelessness and Poverty, *No Safe Place.*

180 *this topic has been circulating through the national conversation since 1999*: Foscarinis, Cunningham-Bowers, and Brown, "Out of Sight—Out of Mind?"

180 *First Amendment right to free speech*: Keyes, "Courts Are Striking Back Against the Criminalization of Homelessness."

180 *Department of Justice filed a brief*: US Department of Justice, "Justice Department Files Brief to Address the Criminalization of Homelessness."

180 *more difficult for municipalities to obtain HUD money if such programs aren't in place*: Pyke, "Local Officials Have Pushed to Criminalize Homelessness for Years. The Feds Are Starting to Push Back."

180 *reduce laws criminalizing homelessness*: Keyes, "Criminalizing Homelessness Can Now Cost Cities Federal Money."

180 *overcrowding in the Pinellas County jail, where many of them were ending up*: Nohlgren and DeCamp, "Tensions Brew over Plans for Homeless Shelter at Pinellas Jail Complex."

181 *slashing the residents' living spaces with box cutters*: Ulferts and Raghunathan, "Police Slash Open Tents to Roust the Homeless."

181 *"excessive arrests of homeless individuals"*: National Law Center on Homelessness and Poverty and National Coalition for the Homeless, *Homes Not Handcuffs*, 11.

181 *"I would rather have them spend money on helping the homeless rather than arresting them"*: Raghunathan, "Public defender will stop working homeless cases in St. Petersburg."

181 *six new ordinances criminalizing homelessness*: National Law Center on Homelessness and Poverty and National Coalition for the Homeless, *Homes Not Handcuffs, 11.*

181 *second meanest city in the country*: Ibid., 33.

181 *"who were routinely penalized for using public space to perform basic bodily functions when they had nowhere else to go"*: Catron, Young, Hargis, et al. v. City of St. Petersburg, US Court of Appeals, Eleventh Circuit.

181 *"Velvet Hammer"*: Delaney, "How a Traveling Consultant Helps America Hide the Homeless."

182 *"perpetuates and increases homelessness through enablement"*: Marbut, "Homeless Needs Assessment and Action Plan for Placer County," 39.

182 *"The mission should no longer be to 'serve' the homeless community"*: Ibid., 21.

182 *ordinances targeting homelessness*: Pinellas County deputy, interview, July 30, 2015.

183 *compared with $125 a day for the jail*: Morrow, "Changes Expected in June Will Address Issues Plaguing Pinellas Safe Harbor Homeless Shelter."

183 *daily average of 450*: Girardi, "Groups Work to Provide Permanent Housing for St. Pete's Homeless."

184 *"not sick enough to qualify for a disability check"*: Rolle, interview, July 17, 2015.

184 *Many substance abuse experts agree with G.W.*: Larimer, et al., "Health Care and Public Service Use and Costs Before and After Provision of Housing for Chronically Homeless Persons with Severe Alcohol Problems."

184 *treating substance abuse alone is inadequate; it must be tackled alongside homelessness simultaneously*: National Coalition for the Homeless, *Substance Abuse and Homelessness.*

184 *put them in a house and wrap services around them, and ensure that they can't lose their house no matter what*: Rolle, interview, November 3, 2015.

184 *housing is a human right and not a privilege*: National Law Center on Homelessness and Poverty, "*Simply Unacceptable.*"

185 *the only effective way of ending homelessness*: Delaney, "How a Traveling Consultant Helps America Hide the Homeless."

186 *two years before getting approved*: Pinellas County deputy, interview, July 30, 2015.

188 *"Right here on the stage"*: Snapp, interview, July 19, 2015.

189 *"Lack of money ain't never stopped me from doing nothing"*: Rolle, interview, July 11, 2015.

190 *It costs $240 to feed them all*: Snapp, interview, July 19, 2015.

190 *"I don't think the congregation's in a mood right now to want to shell out more and more"*: Ibid.

190 *"Everyone consistently says it's the best breakfast in town"*: Rolle, interview, July 17, 2015.

193 *called Marbut back for a follow-up evaluation*: Stanley, "Homeless Issues Again Becoming Problem in St. Petersburg, Consultant Says."

193 *much to the dismay of High Point residents nearby*: Madison, "Neighbors Voice Trouble with Pinellas Safe Harbor Shelter."

193 *5,887 total homeless individuals in Pinellas County*: Young and Moore, *2014 Point-in-Time Housing Survey*, 8.

193 *exactly what they'd counted in 2011*: Pinellas County Homeless Leadership Board, *2011 Point in Time (PIT) Count of Homeless Individuals in Pinellas County*, 10.

193 *From 720 unsheltered people that year*: Ibid.

193 *they were now up to 1,178*: Young and Moore, *2014 Point-in-Time Housing Survey*, 9.

193 *Over 2,500 of those counted were children*: Ibid., 8.

193 *Half of those counted lived in St. Petersburg*: Ibid., 30.

193 *"We've got to step it up"*: Stanley, "Homeless Issues Again Becoming Problem In St. Petersburg, Consultant Says."

193 *delivered his findings to new mayor Rick Kriseman and the city council on June 5*: Marbut, "Follow-Up Review of Homelessness in the City of St. Petersburg."

194 *come to an end the previous October*: Thompson, "Pinellas Authorities Evaluating Homeless Diversion Program."

194 *individuals arrested for minor infractions, such as an open-container charge, were held in solitary confinement*: Smith, interview, January 25, 2016.

195 *"You are asking for an answer to an unanswerable question"*: Thompson, "Pinellas Launching Program to Deal with the Chronically Homeless."

195 *only 93 individuals opted to go to Safe Harbor*: Thompson, "Pinellas Authorities Evaluating Homeless Diversion Program."

195 *jail costs the sheriff's department ten times more*: Stanley, "Officials Once Again Mulling Over Homeless Problem In St. Pete, Pinellas County."

196 *$1.6 million comes out of the sheriff's budget*: Stanley, "Homeless Issues Again Becoming Problem in St. Petersburg, Consultant Says."

196 *those costs are offset by donations from various cities*: Ibid.

196 *a program run by the City of St. Petersburg had bused close to a thousand homeless people, some of them addicted to drugs or mentally ill, across the country*: Frago, "Road to Nowhere."

197 *every homeless veteran by the end of 2015*: US Department of Housing and Urban Development, et al., *Mayors Challenge to End Veteran Homelessness in 2015 Fact Sheet.*

197 *roughly 58,000 veterans were experiencing homelessness in America at the time*: Obama, "The First Lady Announces the Mayors Challenge to End Veteran Homelessness."

197 *In January 2015, Mayor Kriseman of St. Petersburg took up the First Lady's challenge*: Kriseman, "State of the City Address."

197 *up from 5,887 the previous year to 6,853*: Santa Lucia, *Pinellas County Point in Time Homeless Report: 2015.*

197 *including 621 veterans in Pinellas County alone, of the almost 3,000 total in the tri-county area*: Raposa, interview, January 25, 2016.

197 *he tells me the tri-county area is now home to less than two hundred homeless veterans*: Ibid.

201 *"I've made mistakes this week"*: Rolle, interview, August 8, 2015.

202 G.W. *finds a way inside*: *Easy Street*, Wideyed Films.

204 *sleeping on the floor of the Dream Center*: Rolle, interview, November 3, 2016.

204 *"It seemed like they had me in a basement"*: Ibid.

206 *elderly woman who lived alone with her cats*: *Easy Street*, Wideyed Films.

208 *There are similar projects in the works in Oregon, Washington, Wisconsin, Georgia, and Alabama*: Girardi, "Groups Work to Provide Permanent Housing for St. Pete's Homeless."

209 *The 2012 census found that 10.1 percent of American homes—over thirteen million units—were sitting vacant*: Heben, *Tent City Urbanism*, 32.

209 *while maintaining their existing balance of privacy and social interactions*: Ibid., xii.

209 *"my visit to the site . . . left me skeptical"*: Ibid., 106.

209 *Eugene's mayor formed a task force to find "new and innovative solutions"*: Ibid., xi.

210 *gathering yurt, common kitchen, front office, tool shed, and a bathhouse with flush toilets, a shower, and a laundry room*: Ibid., 163.

210 *balances the informal with the formal*: Ibid., 161.

210 *was erected for approximately $100,000*: Ibid., 164.

210 *housing first rather than services*: Girardi, "'Tiny Houses' Help St. Pete Tackle Challenge to House Homeless Veterans."

210 *with a list of city-owned properties for sale*: Pavese, interview, January 30, 2016.

211 *He's going to quit, he reassures me*: Rolle, interview, January 24, 2016.

214 *did not meet HUD criteria for homelessness*: Santa Lucia, *Pinellas County Point in Time Homeless Report: 2015*, 1.

214 *only 3,387 were reported to HUD*: Ibid., 4.

214 *35,900 people were homeless in the state of Florida in 2015*: US Department of Housing and Urban Development, *HUD 2015 Continuum of Care Homeless Assistance Programs Homeless Populations and Subpopulations, State Name: Florida*, 1.

214 *71,446 homeless children alone*: Florida Department of Children and Families, *Council on Homelessness 2015 Annual Report*, 5.

214 *have not been on a lease and had to move more than once in the past sixty days, and will continue to be unhoused due to disabilities or barriers to employment*: Ibid., 6.

215 *or are migrants who haven't yet settled in a new place*: Ibid.

215 *low wages fall further still while housing costs skyrocket*: Stewart, "Obama Will Seek $11 Billion for Homeless Families."

215 *putting $11 billion in federal funds into fighting family homelessness over the next ten years—up from $4.5 billion in 2015*: National Alliance to End Homelessness, *The State of Homelessness in America 2015*, 6.

215 *More than half of homeless families with children live in five states*: Stewart, "Obama Will Seek $11 Billion for Homeless Families."

215 *HUD reported that there were only about 123,000 homeless children nationwide*: Office of Community Planning and Development, US Department of Housing and Urban Development, *The 2015 Annual Homeless Assessment Report (AHAR) to Congress.*

215 *but also that the cost to communities was the same*: Stewart, "Obama Will Seek $11 Billion for Homeless Families."

216 *Molly Coyle, who, according to the G.W. character, calls him "for just about everything"*: Rolle, "Things That Break," 15.

216 *appoints himself the mayor of Williams Park*: Ibid., 10.

216 *"They are young people on the verge of going under"*: Ibid., 15.

216 *He thinks about how they called for Christ after Lazarus died*: Ibid., 17.

216 *He agrees to hold a service for Branna that evening*: Ibid., 45.

216 *"Don't do what you used to do"*: Ibid., 73.

217 *have gone before us to loosen our bonds and show us the way*: Ibid., 74.

217 *calls to tell me he's relapsed again*: Rolle, interview, January 25, 2016.

218 *He's slept through the last three days*: Rolle, interview, January 28, 2016.

218 *"The face of this breakfast might change"*: Rolle, interview, January 30, 2016.

219 *Fifty people crowd inside*: Bolden, interview, January 30, 2016.

219 *cover of the Tampa Tribune*: Girardi, "'Tiny Houses' Help St. Pete Tackle Challenge to House Homeless Veterans."

220 *They'll start building in September*: Bolden, welcoming statements, January 30, 2016.

221 *when she Baker Acted herself*: Megan, interview, January 30, 2016.

223 *rescue dogs for veterans who want them*: Bolden, interview, January 30, 2016.

223 *"so you're ending hunger"*: Ibid.

223 G.W. *texts to say he has something to tell me*: Rolle, text message, February 4, 2016.

224 *the last to be held at Trinity Lutheran*: Rolle, interview, January 30, 2016.

224 *"Hey! I'm writing. How about you?"*: Rolle, text message, February 19, 2016.

Sunshine State

226 *"There have been bad things written about us in the paper"*: Beitl, interview, July 12, 2015.

226 *dragging a broken wing on the side of Gulf Boulevard*: *Flying Free 2*, Matheney Productions.

227 *more than fifty years*: Karas, "At 100, Helen Heath Puts Her Life into Suncoast Seabird Sanctuary."

227 *he lives on the bottom floor*: Guastella, interview, August 3, 2015.

228 *"birds come to grief at the hands of man"*: Matthews, "Volunteers Rescue Injured Wildfowl," 31.

228 *"Heath is at his best saving birds which his fellow citizens have damaged"*: Ibid., 34.

228 *first facility in history to mate the brown pelican in captivity*: McMahon, "Rare Brown Pelican Born at Suncoast Sanctuary."

228 *the bird was endangered and on the verge of extinction*: Cappiello, "Brown Pelicans off the Endangered Species List."

228 *hatched the first brown pelican egg*: Greene, "In a Family Way."

228 *Dewar's Scotch had featured Ralph in their Dewar's Profile ad campaign*: Dewar's, "Dewar's Profiles: Ralph Heath."

229 20/20 *did a special on Ralph and the sanctuary then the* Today *show*: Lindberg, "Suncoast Seabird Sanctuary No Longer Taking in Injured Birds."

229 *the New York Times*: Lemkowitz, "A Sanctuary for Seabirds in Florida."

229 *"Birds: Our Responsibility"*: "Birds: Our Responsibility," *St. Petersburg Times.*

229 *Disney came to the sanctuary seeking animal stars for its Discovery Island*: Mitchell, "Good Birds Make Good Neighbors."

229 *zoos all over the world, in Greece*: Guastella, interview, August 3, 2015.

229 *Singapore, Spain*: Williams, "World-Renowned Sanctuary Most Definitely for the Birds."

229 *and Barbados*: Miller, "Pelican Venture."

229 *"all atilt, with no tether"*: Heath, interview, November 4, 2015.

229 *took him bird-watching in his amphibian aircraft*: Associated Press, "Florida Pelicans Retire in Texas."

229 *he shot films from the sanctuary yacht, the* Whisker: Guastella, interview, August 3, 2015.

229 *Ralph and Linda divorced three years after the sanctuary opened*: Divorce certificate, Sixth Judicial Circuit of Florida.

229 *in 1982 he married Beatrice Busch*: Evertz, "300-Acre Estate Is Setting for Busch-Heath Wedding."

229 *Beatrice had spent winters in Tampa Bay with her beer magnate father*: Ibid.

230 *had just finished filming a television special about the migration of humpback whales with her filmmaker ex-husband*: "Beatrice Busch Promotes Sanctuary," *Boca Beacon.*

230 *She arrived just as Ralph was leaving to rescue a pelican hooked by a fisherman*: Ibid.

230 *grabbing birds with their hands, getting nipped by cormorants*: Miller, "Beatrice Busch-Keefe."

230 *Grant's Farm, a three-hundred-acre St. Louis estate*: Evertz, "300-Acre Estate Is Setting for Busch-Heath Wedding."

230 *chimpanzees and elephants Gussie had trained himself*: Thomas, "August A. Busch, Jr. Dies at 90; Built Largest Brewing Company."

230 *Within three years of marrying Ralph, Beatrice had given birth to three sons*: Guastella, interview, August 3, 2016.

230 *Within another two years, Ralph and Beatrice had separated*: Marriage certificate, Commonwealth of Virginia Department of Health, Division of Vital Records.

230 *Ralph refused to go with her. He couldn't leave his birds*: Heath, interview, November 4, 2015.

230 *In the beginning, the birds had united Ralph and Beatrice*: "Beatrice Busch Promotes Sanctuary," *Boca Beaton.*

230 *they postponed their trip until the St. Petersburg City Council reached a verdict*: Evertz, "300-Acre Estate Is Setting for Busch-Heath Wedding."

231 *It was a plan he'd been working on since 1974*: "Council Delays Gateway," *St. Petersburg Evening Independent.*

231 *buying land as a way to preserve it against booming development along Florida's coasts*: Griffin, "Andrew's Island—The Third One."

231 *had community support*: East, "'Nature Lovers' Plan Petition Drive for Bird Sanctuary Land Sale."

231 *the city wanted to zone the land for industrial use*: "Council Says It's

Willing to Meet Nov. 10 with Seabird Sanctuary Founder," *St. Petersburg Times.*

231 *They were expected to come to a decision in December*: Stiff, "City's Stand on Sanctuary Is for the Birds."

231 *petitions in favor of the park collected hundreds of names*: Ibid.

231 *"These individuals do not yet realize that mankind and wildlife live in the same ecosystem"*: Meinhardt, "'Double-Cross' Allegations over Site for Sanctuary."

231 *In the end, the city turned the plot into a landfill*: East, "Sod Farm Land to Be Considered for Landfill Use."

231 *the largest nonprofit wild bird hospital in the country*: Lindberg, "Suncoast Seabird Sanctuary No Longer Taking in Injured Birds."

231 *published a quarterly newsletter,* Fly Free: Suncoast Seabird Sanctuary, *Fly Free*, Summer 2008.

232 *At its largest the sanctuary employed forty-four workers*: Vaughan and Vaughan, interview, August 10, 2015.

232 *internationally praised bird rehabilitator, Barbara Suto*: Goff, "New Bird Rescue Operation Fills a Void."

232 *in state-of-the-art emergency facilities and a surgical center*: *Flying Free 2*, Matheney Productions.

232 *licensed veterinary assistants*: Godwin, interview, July 24, 2015.

232 *groundskeepers; office staff; a marketing director*: Vaughan and Vaughan, interview, August 10, 2015.

232 *all-volunteer rescue team bringing in up to forty-five new birds every day*: Suncoast Seabird Sanctuary, "Our Birds . . . By the Numbers."

232 *volunteers keeping tight shifts on the grounds and in the sanctuary hospital*: Walls, interview, July 14, 2015.

232 *Chris was one of only three people there*: Ibid.

233 *There were now six paid employees working at the sanctuary*: Ibid.

233 *People were steadily leaving jobs at the sanctuary*: Douglas, "Seabird Sanctuary Workers Say Donations Disappeared."

233 *Ralph was accused of stealing money out of the donation boxes—outed to the press by former employees*: Ibid.

233 *Ralph announced the sanctuary would no longer be rescuing birds*: Lindberg, "Suncoast Seabird Sanctuary No Longer Taking in Injured Birds."

233 *transitioning to a volunteer-based staff due to its inability to keep up*

with debts: Douglas, "Suncoast Seabird Sanctuary Must Pay $21,336 in Back Wages."

233 *The IRS was pursuing them for almost $200,000 in unpaid payroll taxes*: Lindberg, "State Files Tax Lien Against Troubled Suncoast Seabird Sanctuary."

233 *They owed more than $21,000 in employee back pay*: Douglas, "Suncoast Seabird Sanctuary Must Pay $21,336 in Back Wages."

233 *Ralph was arrested and charged with workers' compensation fraud*: Lindberg, "Suncoast Seabird Sanctuary Founder Charged with Workers' Compensation Fraud."

233 *a man had slipped in a truck while delivering ice at the sanctuary*: Eslick, interview, November 4, 2015.

233 *"It was the management"*: Walls, interview, July 14, 2015.

234 *"Then they'd go off and die, and not bother the chickens"*: *Flying Free 2*, Matheney Productions.

234 *Licensed rehabbers pick birds up from the hospital a few times a week so they can receive care the sanctuary can't provide*: Walls, interview, July 14, 2015.

234 *She's a former kindergarten teacher*: Vickery, interview, July 12, 2015.

235 *sanctuary volunteers used to go through special training to feed baby birds*: Suncoast Seabird Sanctuary, "Volunteers."

235 *"He's an OCD. Hoarder"*: Kraut and Vaughan, interview, July 29, 2015.

236 *Ralph signed over the deeds for the sanctuary property and his house to Seaside Land Investments LLC*: Lindberg, "Suncoast Seabird Sanctuary No Longer Taking in Injured Birds."

236 *heirs to the Anheuser-Busch fortune and the fortune of their stepfather*: "All the Kings' Children," *St. Louis Magazine*.

236 *"It's wall-to-wall pigeons"*: Kraut and Vaughan, interview, July 29, 2015.

236 *can't legally release it back into the wild*: "Wildlife Rehabilitation Permit," Rule 68A-9.006.

236 *"Well, yeah. He comes in and looks around"*: Kraut and Vaughan, interview, July 29, 2015.

236 *six hundred pounds of fish*: Suncoast Seabird Sanctuary, "Our Birds . . . By the Numbers."

237 *Greg Vaughan was working at the sanctuary when the Deepwater Horizon oil spill happened*: Vaughan, interview, July 12, 2015.

237 *claiming there's too much paperwork involved*: Vaughan and Vaughan, interview, August 10, 2015.

237 *some are quite large, whether monetary or in the form of cars and yachts*: Guastella, interview, August 3, 2015.

237 *"the sanctuary's troubles would be over, for a time at least"*: Bothwell, "Seabird Sanctuary Gains Recognition—but the Bills!"

237 *Every few months, someone died and left millions of dollars to the sanctuary*: Guastella, interview, August 3, 2015.

237 *Ralph's mother, Helen, managed the business*: Vaughan and Vaughan, interview, August 10, 2015.

237 *she fell ill and Micki took over*: Eslick, interview, November 4, 2015.

238 *Jimbo's band held a benefit concert for the sanctuary down on St. Pete Beach*: Guastella, interview, August 3, 2015.

238 *"I had hair down to here"*: Ibid.

238 *he'd never been interested in the sanctuary's operations*: Ibid.

238 *With his mother running the office, he never thought to learn how it ran*: Vaughan and Vaughan, interview, August 10, 2015.

238 *When publicity went south, so did donations*: Lindberg, "Suncoast Seabird Sanctuary No Longer Taking in Injured Birds."

238 *he said, wrongfully, that birds faced famine*: Lindberg, "Seabird Sanctuary Founder Sued Twice, Accused of Selling Car to Two Different Buyers."

238 *his waterfront home, which he'd purchased for $300,000, the sanctuary yacht, which he'd purchased for $355,000*: Lindberg, "Suncoast Seabird Sanctuary Warehouse to Be Auctioned to Pay Debt."

239 *claimed he was using it to research the effects of plastic pollution on the ocean floor*: Ibid.

239 *"I came home with three thousand dollars in my pocket"*: Guastella, interview, August 3, 2015.

239 *Greg spotted an ad for a groundskeeper in the paper*: Vaughan, interview, July 12, 2015.

239 *Kellie, a certified nurse, moved in with her*: Vaughan and Vaughan, interview, August 10, 2015.

239 *Greg told me that Ralph owes him and Kellie each $7,000*: Vaughan and Vaughan, interview, July 30, 2015.

239 *"sort of a consultant"*: Vickery, interview, July 12, 2015.

240 *crippled Kellie's ability to maintain donor relations*: Vaughan and Vaughan, interview, July 30, 2015.

240 *attempting to sell the same 1963 Corvette Stingray to two different buyers*: Lindberg, "Seabird Sanctuary Founder Sued Twice, Accused of Selling Car to Two Different Buyers."

240 *"Ralph's an idiot"*: Vaughan and Vaughan, interview, July 30, 2015.

240 *a man had died and left his estate to the sanctuary*: Vaughan and Vaughan, interview, August 10, 2015.

240 *"The thing is, he's not doing it for the sanctuary"*: Ibid.

241 *there was a foreclosure suit against it back in 2013*: Lindberg, "Foreclosure Suit Filed against Suncoast Seabird Sanctuary Property."

241 *the warehouse was put up for auction*: Lindberg, "Suncoast Seabird Sanctuary Warehouse to Be Auctioned to Pay Debt."

242 *At the last minute, the auction was canceled*: Cooper vs. Suncoast Seabird Sanctuary, et al.

242 *Ralph's sons registered an LLC named after the Starkey Road address*: "12388 Starkey Road, LLC," Corporation Wiki.

242 *paid for the warehouse*: Vaughan and Vaughan, interview, August 10, 2015.

242 *"You don't want to know"*: Walls, interview, July 12, 2015.

242 *35,000 square feet, according to News Channel 8*: Douglas, "Foreclosure Filed over Seabird Sanctuary Warehouse."

242 *has two floors, according to Jimbo*: Guastella, interview, August 3, 2015.

242 *Ralph refuses to believe what people say about him is true*: Vaughan and Vaughan, interview, July 12, 2015.

242 *"The media, as you well know, likes sensationalism"*: Heath, interview, November 4, 2015.

242 *"He has to win"*: Guastella, interview, August 3, 2015.

243 *He replied that he had one hundred pigeons*: Heath, interview, July 12, 2015.

243 *"We got a lot of baby birds in the hospital last night"*: Heath, interview, August 1, 2015.

244 *"Did I miss a call from you this morning?"*: Vaughan, interview, August 3, 2015.

245 *"They want to keep it alive and they know it won't stay alive with him involved"*: Guastella, interview, August 3, 2015.

245 *some falling in love, as Ralph had described*: Heath, interview, July 30, 2015.

245 *some sick and waiting to die until he returned home*: Heath, interview, August 12, 2015.

247 *von Gontard estate, Oxbow Farm*: Orcutt, "Down on the Farm: Tours Aim to Provide Missing Piece to Story."

248 *Andrew is a commercial airline pilot based in St. Louis, has a law degree, and is making his way toward being a judge*: Guastella, interview, August 3, 2015.

248 *recently launched an upscale bar called the Eberhard*: Peppard, "Dallas Bar the Eberhard from Brothers with Beer in the Family."

248 *as well as a dating app called Courtem*: Hall, "Anheuser-Busch Brothers Tap into Dating App Market with Support from Mark Cuban."

248 *"this is where the huge fear is going on here right now"*: Guastella, interview, August 3, 2015.

249 *"He's just made a lot of stupid mistakes"*: Vaughan and Vaughan, interview, August 10, 2015.

250 *sitting under three feet of bird shit, infested with bees*: Guastella, interview, August 3, 2015.

251 *The sanctuary's score on Charity Navigator*: "Suncoast Seabird Sanctuary: Historical Ratings," Charity Navigator.

251 *purchased by the sanctuary back in 1997, from a funeral director*: Deed of sale from Zeta G. Bobbitt to Suncoast Seabird Sanctuary Inc.

251 *generate revenue for the sanctuary by renting out the house*: Eslick, interview, November 4, 2015.

251 *Micki had retired from Verizon after thirty-four years*: Ibid.

251 *doing the sanctuary's books by hand for almost forty years*: Karas, "At 100, Helen Heath Puts Her Life into the Suncoast Seabird Sanctuary."

251 *And in 2007, she asked Micki to help her with administrative duties*: Eslick, interview, November 4, 2015.

251 *Micki's husband, Slick, was against the transition*: Ibid.

252 *Michelle Glean Simoneau was the sanctuary's marketing and public relations director*: Mullane Estrada, "Financial Problems Strain Suncoast Seabird Sanctuary."

252 *keeping on general good terms with the public*: Vaughan and Vaughan, interview, August 10, 2015.

252 *built the sanctuary back up to near its former glory*: "Suncoast Seabird Sanctuary: Historical Ratings," Charity Navigator.

252 *Ralph promised Michelle she'd be executive director when he retired*: Vaughan and Vaughan, interview, August 10, 2015.

252 *"Ralph let her have it because it's his first cousin"*: Ibid.

252 *Ralph gave Micki the new title of operations manager*: Ibid.

252 *"I would say we need to use this money for what it was intended for"*: Douglas, "Suncoast Seabird Sanctuary Finances Face Review."

252 *she was weeks behind in pay*: Ibid.

252 *"Micki hired eight or nine more employees, including her son"*: Vaughan and Vaughan, interview, August 10, 2015.

253 *sanctuary had allowed employees' health insurance to lapse without telling them*: Lindberg, "U.S. Labor Department Investigating Suncoast Seabird Sanctuary."

253 *Kellie began working full-time at a nursing home to make ends meet*: Vaughan and Vaughan, interview, August 10, 2015.

253 *"Mrs. Heath, to me, was like my grandma"*: Ibid.

255 *"We were never able to verify another account"*: Heath, interview, November 4, 2015.

255 *supposedly for health reasons*: Vaughan, interview, October 28, 2015.

255 *they were prepared to quit if Ralph didn't leave*: Vaughan and Vaughan, interview, August 10, 2015.

255 *Instead I received a response from Adrianne*: Beitl, interview, October 28, 2015.

255 *"We never had any proof"*: Heath, interview, November 2, 2015.

257 *"They bought the mortgage and took it over"*: Lindberg, "Mistake Gives Suncoast Seabird Sanctuary Owner Years of Tax Breaks."

257 *Beatrice and Ralph's sons had given him two weeks to move*: Vaughan and Vaughan, interview, August 10, 2015.

258 *"reasons that are not clear but appear to be nefarious"*: Doty, letter about Ralph Heath.

259 *Micki's friend paid $3,000 for the cremation*: Eslick, interview, November 4, 2015.

259 *they filed for Chapter 13 bankruptcy*: Notice of bankruptcy case filing, US Bankruptcy Court, Middle District of Florida.

259 *"I didn't do any embezzling"*: Eslick, interview, November 4, 2015.

259 *they were much like siblings, always very close*: Ibid.

259 *heated pools for the turtles so they wouldn't be cold in the winter*: Heath, interview, November 4, 2015.

260 *"Why don't you go ask Ralph how much money he's embezzled from the sanctuary?"*: Eslick, interview, November 4, 2015.

260 *Omar Bsaies, a sanctuary board member*: Suncoast Seabird Sanctuary, *Fly Free*, Spring/Summer 2012.

260 *former US diplomat now living in the Philippines*: San Diego, "Ali's Wonderland a Labor of Love."

261 *a time when the sanctuary needed serious help*: Eslick, interview, November 4, 2015.

261 *"he runs over and kind of ruins it"*: Guastella, interview, November 27, 2015.

261 *In 1986, a woman named Anne Bywater had inherited four hundred acres of land*: Wiggins, interview, November 24, 2015.

262 *The land was mostly underwater and was environmentally protected*: Holan, "Pasco County 400-Acre Seabird Sanctuary for Sale."

262 *his plans for the zoological park had been annihilated by the no-name storm of 1993*: Ibid.

262 *"I honestly feel that Ralph Heath always had grand ideas that he never carried through"*: Wiggins, interview, November 24, 2015.

262 *"You don't do business like that"*: Eslick, interview, November 4, 2015.

262 *selling Ralph's house would be in the sanctuary's best interest*: Ibid.

263 *"what it takes to keep the sanctuary alive and well"*: Marler, interview, November 6, 2015.

263 *it was the first fuel-injected Corvette to enter the state of Florida*: Lindberg, "Seabird Sanctuary Founder Sued Twice, Accused of Selling Car to Two Different Buyers."

263 *"Then you have the '61 and the '63 to go along with your other eighteen or twenty cars"*: Heath, interview, November 4, 2015.

264 *"only triple red un-restored and all original 1963 Corvette of its kind"*: Lindberg, "Seabird Sanctuary Founder Sued Twice, Accused of Selling Car to Two Different Buyers."

265 *"they're a wildlife object"*: Heath, interview, August 1, 2015.

265 *"I have to swipe away some spider webs to get out of my room"*: Guastella, interview, November 27, 2015.

266 *Tara Gallagher, a former girlfriend of Ralph's and employee of the sanctuary*: "Memorial for Suncoast Seabird Sanctuary Employee Set for Sunday," *Tampa Bay Times*.

266 *It's not uncommon for Ralph to forage in the sanctuary's trash cans*: Guastella, interview, November 27, 2015.

267 *the story of Grungy, the orphaned pigeon*: Heath, interview, August 1, 2015.

268 *Birds, especially pigeons, are linked to a slew of human illnesses*: Nordqvist, "Birds and Their Droppings Can Carry over 60 Diseases."

268 *Bird fancier's lung is one of the most common*: "Health Guide: Hypersensitivity Pneumonitis," *New York Times*.

268 *Other bird-linked illnesses, such as cryptococcal meningitis, affect the brain*: "Health Guide: Meningitis-Cryptococcal," *New York Times*.

268 *spinal tap showed it wasn't that, but rather cryptococcal meningitis*: Eslick, interview, November 4, 2015.

268 *"Birds are dying at an alarming rate"*: Vergara, email to von Gontard brothers.

268 *"Wings of Change: A Win-Win for All"*: Taylor, "Save the Suncoast Seabird Sanctuary 2012."

268 *the sanctuary had received a $100,000 donation from the Embassy of Qatar*: Mullane Estrada, "Suncoast Seabird Sanctuary on Indian Shores Gets $100,000 Surprise from Qatar."

269 *there was no warehouse that could accommodate birds*: Eslick, email to Robin Vergara, June 14, 2012.

269 *A new warehouse never materialized*: Vergara, interview, November 10, 2015.

269 *"it would have been a disaster like no other"*: Ibid.

269 *Fed up, Robin finally left to open his own bird sanctuary*: Vergara, interview, July 24, 2015.

269 *In April 2014, Florida Fish and Wildlife carried out an inspection of the Suncoast Seabird Sanctuary, Ralph's house, and the warehouse*: Florida Fish and Wildlife Conservation Commission, inspection report, April 30, 2014.

269 *"confined in unsanitary conditions"*: Meacham, "Authorities Issue 59 Violations against Suncoast Seabird Sanctuary."

270 *"no longer rehabbing birds at his home or the warehouse"*: Florida Fish and Wildlife Conservation Commission, inspection report, June 17, 2014.

270 *There is no regulation on the number of pigeons a person can have*: Gregory, interview, November 29, 2015.

270 *In January 2016, Jimbo told me, Andrew, Alex, and Peter planned to return to Florida to force Ralph out of his house and into the warehouse*: Guastella, interview, November 27, 2015.

271 *"one thing that I've always worried about is hurricanes"*: Heath, interview, November 4, 2015.

273 *Chief Jim Billie is the longest-serving chairman of the Seminole tribe of Florida*: Gallagher, "The Rise and Fall of Chief Jim Billie."

276 *In January 2016, Tampa Bay Newspapers ran an online story about a blue heron approaching Ralph at the sanctuary*: Goff, "Injured heron finds its way back to seabird sanctuary."

276 *any animals kept by the sanctuary as permanent residents must be made available for public viewing*: "Wildlife Rehabilitation Permit," Rule 68A-9.006.

277 *Lieutenant Steve Delacure and Officer Robert O'Horo investigated the claim*: Douglas, "Photos Show 'Deplorable' Conditions for Wildlife Held Captive by Suncoast Seabird Sanctuary Founder."

277 *"A cloud of rancid-smelling particulate hung in the air"*: Florida Fish and Wildlife Conservation Commission, inspection report, May 5, 2016.

277 *The wing of a federally protected laughing gull had been amputated and bound in duct tape*: Ibid.

278 *He pled not guilty to all charges*: Douglas, "Embattled Suncoast Seabird Sanctuary Founder Pleads Not Guilty."

278 *"My assistant who was working there passed away unexpectedly"*: Heath, interview, June 5, 2016.

ABOUT THE AUTHOR

SARAH GERARD is the author of the novel *Binary Star* (Two Dollar Radio), which was a finalist for the Los Angeles Times Award for First Fiction and appeared on national Book of the Year lists. Her short stories, essays, interviews, and criticism have appeared in the *New York Times*, *Paris Review Daily*, *Los Angeles Review of Books*, *Granta*, *Bookforum*, *Joyland*, *Vice*, *BOMB*, and other journals, as well as in anthologies for *Joyland* and the *Saturday Evening Post*. She's been supported by grants and fellowships from Yaddo, Tin House, and PlatteForum. She writes a monthly column for *Hazlitt* and teaches writing in New York City.